Arthur Milnes Marshall

Lectures on the Darwinian Theory

Arthur Milnes Marshall

Lectures on the Darwinian Theory

ISBN/EAN: 9783743424074

Manufactured in Europe, USA, Canada, Australia, Japa

Cover: Foto ©berggeist007 / pixelio.de

Manufactured and distributed by brebook publishing software (www.brebook.com)

Arthur Milnes Marshall

Lectures on the Darwinian Theory

THE DARWINIAN THEORY

By the same Author.

THE FROG.

An Introduction to Anatomy, Histology and Embryology.
Fifth Edition. 1894. 4s.
MANCHESTER: J. E. CORNISH.

A JUNIOR COURSE OF PRACTICAL ZOOLOGY.

In conjunction with C. HERBERT HURST, Ph.D.,
Demonstrator and Assistant-Lecturer in
Zoology at Owens College.
Third Edition. 1892. 10s. 6d.

VERTEBRATE EMBRYOLOGY.

A Text-book for Students and Practitioners.
First Edition. 1893. 21s.
LONDON: SMITH, ELDER & CO., 15 WATERLOO PLACE.

BIOLOGICAL LECTURES AND ADDRESSES.

Edited by C. F. MARSHALL, M.D., B.Sc., F.R.C.S.
First Edition. 1894. 6s.
LONDON: DAVID NUTT 270-271, STRAND.

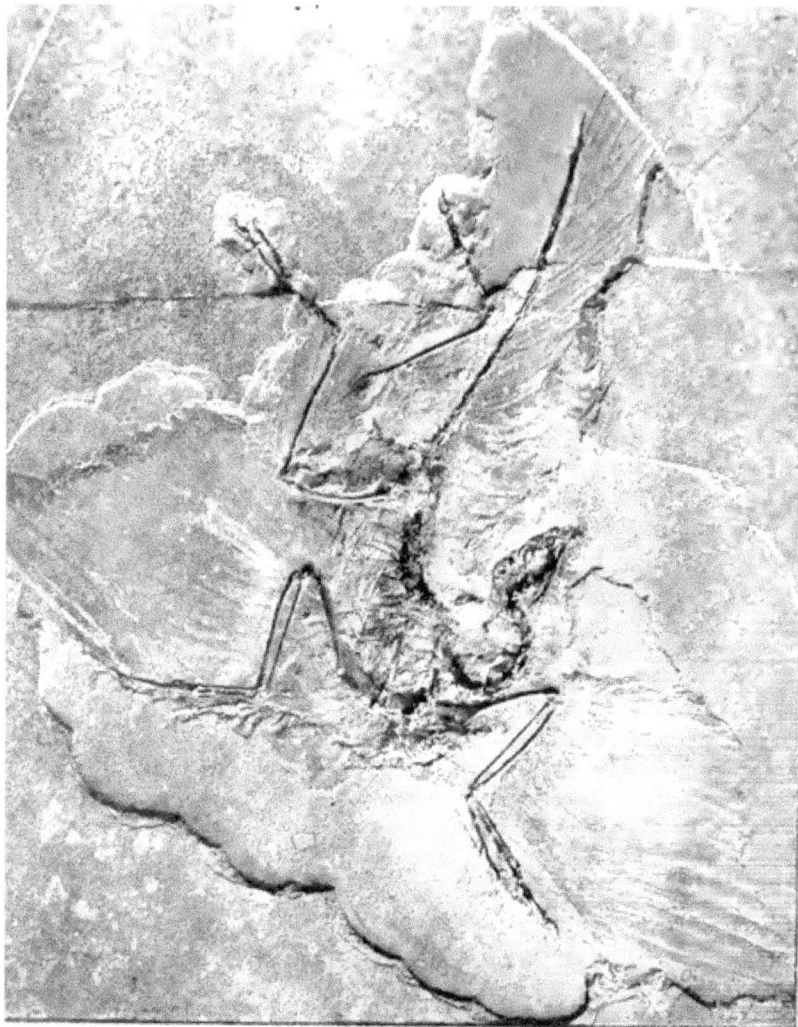

ARCHÆOPTERYX

LECTURES

ON THE

DARWINIAN THEORY

DELIVERED BY THE LATE

ARTHUR MILNES MARSHALL.

M.A., M.D., D.Sc., F.R.S.

PROFESSOR OF ZOOLOGY IN OWENS COLLEGE;
FORMERLY FELLOW OF ST. JOHN'S COLLEGE, CAMBRIDGE.

EDITED BY

C. F. MARSHALL

M.D., B.Sc., F.R.C.S.

WITH 37 ILLUSTRATIONS
MOSTLY FROM ORIGINAL DRAWINGS AND PHOTOGRAPHS

LONDON
DAVID NUTT, 270–271, STRAND
1894

Printed by BALLANTYNE, HANSON & CO.
At the Ballantyne Press

PREFACE

THIS volume consists of a series of lectures delivered by the late Professor Marshall in connection with the Extension Lectures of the Victoria University during the year 1893. These have been amplified in some places where occasion required from other unpublished lectures by the same author.

Owing to the variable nature of the original MS., some parts being fully written, while others were in the form of notes, there is a consequent variability in the fulness of the text, and any errors or discrepancies should be attributed to the editor and not the author.

It is greatly to be regretted that these Lectures were not elaborated and prepared for publication by the author himself; but, in spite of the shortcomings of the book, I trust that it may form a useful contribution to the literature of Darwinism, since the Lectures were delivered by one of Darwin's most earnest disciples.

The large majority of the illustrations are taken from original drawings by the author, or from photographs from nature. Some of the drawings are modified from other sources, and in these cases the source has been acknowledged. The blocks for the illustrations have been prepared with great care by Messrs. Walker & Boutall.

In preparing this book I have received valuable assistance from Mr. W. E. Hoyle, Dr. C. H. Hurst, Professors Hickson and Herdman, and Mr. W. Garstang. I must also thank Mr. Francis Darwin for permission to reproduce the extracts in Lecture VIII.

C. F. M.

London. *October 1894.*

CONTENTS

LECTURE I. (Pp. 1–24.)

HISTORY OF THE THEORY OF EVOLUTION.

LECTURE V. (Pp. 116–150.)

THE COLOURS OF ANIMALS AND OF PLANTS.

Colour may be non-significant, as in the case of the redness of the blood of many animals, or the colours of animals living in darkness. More usually, however, colour can be shown to have a direct relation to the welfare of the individual or species, and to be attributable to the action of Natural Selection.

Of late years our knowledge on this subject has advanced greatly, mainly through the observations of Mr. Wallace, supplemented by those of Mr. Bates, Mr. Trimen, Mr. Poulton, and others.

COLOURS OF ANIMALS.

1. *Apatetic Coloration.*

Apatetic Coloration serves to hinder recognition : it may be considered under three heads :—

(*a*) *Protective resemblances :* aiding escape from enemies. The resemblances are usually either to plants, as in the case of the leaf insects, and stick insects, and of the green coloured caterpillars and other frequenters of plants or trees ; or else to inanimate objects, as in the case of the whiteness of the Arctic Hare, and of other defenceless Arctic animals.

A peculiar and interesting form of protective resemblance is afforded by the cases of mimicry, in which a defenceless butterfly or other animal escapes attack through its superficial resemblance to a noxious or venomous animal.

(*b*) *Aggressive resemblances* are cases in which the object gained, like that of the wolf in sheep's clothing, is to facilitate approach to the prey through a superficial resemblance to other objects. Examples are afforded by the whiteness of the Polar Bear and other predacious Arctic animals, or the colouring of the Lion or Tiger.

(*c*) *Alluring resemblances* are cases in which the coloration is such as to cause the animal to resemble a flower or other attractive object, and so to entice the approach of prey.

2. *Sematic Coloration.*

Sematic Coloration is the direct opposite of Apatetic Coloration. and aims at securing recognition; it is of two chief kinds :—

(*a*) *Warning Colours.* Insects or other animals which are inedible owing to an unpleasant flavour or other cause, are usually very conspicuously coloured : the object being to advertise their inedibility, and to secure instant recognition, lest they should be killed by mistake.

(*b*) *Recognition Markings*, such as the white tail of the Rabbit and the markings on certain Deer, are believed to aid recognition by members of the same species.

3. *Epigamic Coloration.*

Under this head all the cases of Sexual Coloration are included, in which, as in the Peacock, the bright colouring is confined to one sex, or is at any rate more marked in it, and is displayed for the purpose of attracting the opposite sex.

COLOURS OF PLANTS.

The bright colours of Flowers and of Fruits serve to attract the insects which fertilise the flowers, and birds and mammals which secure the dispersal of the seeds.

Cross-Fertilisation. The methods of ensuring cross-fertilisation in *Orchids.*

Attractive, and Protective Fruits.

LECTURE VI. (Pp. 151–172.)

OBJECTIONS TO THE DARWINIAN THEORY.

Natural Selection explains the present structure of animals as due to the slow accumulation of small variations, which are transmitted by inheritance from generation to generation. Natural Selection acts primarily for the good of the species, not of the individual.

THE DESCENT OF MAN.

Man distinctly an animal; moreover a Vertebrate and a Mammal. Comparison of the structure and development of man with other animals. Points of resemblance of man to the higher apes. Rudimentary organs present in man. The tendency to Reversion.

The few structures found in man which are peculiar to him.

Other characteristics of man compared with lower animals.

Application of the Darwinian Theory to the Language of Man.

Conclusion. Causes known which account for the structure, language, and habits of man. The possibility of other attributes not explained by these laws.

LECTURE VIII. (Pp. 200—228.)

THE LIFE AND WORK OF DARWIN.

SUMMARY OF LEADING EVENTS.

Birth . . . . 12th February, 1809.
At School at Shrewsbury . . . . . 1818—1825.
At University of Edinburgh . . . . . 1825—1827.
At University of Cambridge . . . . . 1828—1831.
Voyage of H.M.S. *Beagle* . . . . . . 1831—1836.
First Note-Book on "Origin of Species" commenced 1st July, 1837.
Marriage . . . . 29th January, 1839.
"Journal of Researches" published . . . . . 1839.
"Zoology and Geology of Voyage of *Beagle*" . . 1840—1846.
First outline of "Origin of Species" written . . . . 1842.
"Monograph on the Cirripedia" . . . . . 1846—1854.
Announcement of the Theory of Natural Selection } by Darwin and Wallace . . . . . } 1st July, 1858.
"Origin of Species" published . . . . . 1859.
"Fertilisation of Orchids" . . . . . . 1862.
"Variation of Animals and Plants under Domestication" . 1868.
"Descent of Man" . . . . 1871.
"Expression of the Emotions" . . . . . 1872.
"Movements and Habits of Climbing Plants" . . . 1875.
"Insectivorous Plants" . . . . . . 1875.
"Cross and Self-Fertilisation in Plants" . . . 1876.
"The different Forms of Flowers in Plants of the same species" . . . . 1877.

LIST OF ILLUSTRATIONS

THE DARWINIAN THEORY

LECTURE I

HISTORY OF THE THEORY OF EVOLUTION

VERY early in human history the necessity arose for collective names to indicate the various groups and kinds of animals; to distinguish those that were useful for food, clothing, or weapons, &c., from those which were useless or dangerous. An early mode of classification was according to habitat; thus Solomon divided animals into beasts, fowls, creeping things, and fishes. This classification was hardly improved upon till the time of ARISTOTLE, 384–322 (?) B.C., a man for whose intellectual power the word stupendous seems barely adequate. Aristotle made many shrewd and acute observations which were not understood at the time, and were redis-covered 2000 years later: he was a man far ahead of his age.

After this follows a great gap in the history. Facts were steadily accumulating, but there was no system or governing principle. Nothing was known of the history of life on earth, and there was indeed no idea that there had been such a history. Scattered facts imperfectly ascertained were mixed with much superstition, and associated rather with witchcraft than with science.

LINNÆUS, 1707–1778, was the founder of modern scientific natural history. He was a self-made man, and had a long-continued struggle with poverty during the early part of his career. A botanist from his birth, he was never tired of learning facts about plants. Linnæus was originally intended for the Church, but got on so badly at the schools that in 1727 his father proposed to apprentice him to a shoemaker; however, a friend interested in the young botanist persuaded the father, a poor pastor, to let him learn medicine. With this object in view Linnæus went to Lund, and in 1728 to Upsala. In 1732 he was sent by the Literary and Scientific Society of Upsala to Lapland. In 1735 he went to Holland, and was introduced to Boerhaave, the celebrated physician of Leyden, and by him was introduced to a rich banker, who became his patron and in 1736 sent him to England. He afterwards went to Paris and eventually to Stockholm, where he gained his living by practising as a physician. In 1741 he was appointed Professor of Medicine at Upsala, and at the end of the year exchanged the chair of Medicine for that of Botany and Natural History. His poverty was now over and his fame well established.

Linnæus was a man of extraordinary industry, and sent out his pupils in all directions, thereby collecting information and specimens from every part

of the world, with the minute description, arrangement, and classification of which he charged himself.
The first edition of his " Systema Naturæ " appeared
in 1735, and the twelfth edition in 1766. The great
merit of Linnæus lay in the fact that he was a man
of method, and his strength lay in the orderly
arrangement of his knowledge. He introduced a
clear, precise, and definite terminology, and that this
might be universal he employed Latin, the universal
scientific language even at the present day. Still
more important than this, he limited and defined the
use, not only of words but of names, and established
the binomial system. For instance, all roses were
called Rosa : the common dog-rose was described as
Rosa sylvestris vulgaris, flore odorata incarnato,
which means, " The common rose of the woods, with
a flesh-coloured sweet-scented flower." This system
was sufficient perhaps, but clumsy. As we with
our friends find it convenient to use a double name,
so with plants and animals Linnæus gave a double
name—a *generic* and a *specific*, or, as he called it, a
"trivial" name. Larger groups of families and
orders were defined by a few easily recognised
points, such as the number of stamens and pistils
in the case of flowers, and the mode of their
attachment.

This precise and accurate terminology and orderly
arrangement for the first time made it possible to
determine the number of different kinds of animals
and plants, and to define their characters. This was
a service of the utmost importance, and through
Linnæus botany first became a science ; so also with

zoology, though for a time less perfectly than botany. For the first time it became possible to think or speak of the animal or vegetable kingdom comprehensively, and when a name was used, to know precisely what it signified. This greater exactness at once brought men face to face with the problem— What is a species? What do we mean by species of animals or plants?

I will illustrate this by examples of what Linnæus meant by species. For instance, the jackdaw, the raven, the rook, and the crow, are all species of the genus *Corvus*. These birds are clearly more like one another than either of them is like a starling or an eagle; in other words, there is a certain resemblance or affinity between them. Yet, they always differ in certain slight peculiarities of structure, form, and habits; also, they always produce their own kind, and do not interbreed—at any rate, as a rule. The jackdaw (*Corvus monedula*) is grey at the sides of its neck, and builds its nest in holes or cavities in rocks, churches, chimneys, and uninhabited houses. It feeds chiefly on insects, and is much the smallest of the four birds. The raven (*Corvus corax*) is the largest of the four, and is black all over. It makes bulky nests on crags or in trees. The rook (*Corvus frugilegus*) is a trifle smaller than the crow, and lives in noisy flocks. It is black, with a grey forehead and throat. Its nests are found in small trees, often near human habitations. The crow (*Corvus corone*) is smaller than the raven, and is black, tinged with green on the neck and throat, and purple on the back. So also the lion, tiger, leopard, and domestic cat are all species of one genus, differing in points of structure, size, and habits. Again, the pine, fir, and larch are all species of the genus *Pinus*.

The idea of a species is therefore one separated by distinctive characters, which are constant and reproduced in the offspring. Jackdaws do not lay eggs from which crows are hatched, nor do cats give birth to lions. Linnæus' idea of species was that they always had been distinct; that at the original stocking of the earth one pair of each kind or species was created, and that the existing species of animals and plants are the direct descendants of these original inhabitants. The initial objection to this view is that, if there were only one pair of each species to start with, they would immediately have eaten one another up, or have themselves died; herbivorous animals devouring plants, and being devoured in their turn by carnivorous animals. Linnæus does not seem to have troubled himself much with the problem, and appears to have adopted the current views of the day without inquiry. The case was, however, different with some of his contemporaries.

BUFFON, 1707–1788, was a contemporary of Linnæus, but a man of very different stamp. He came of a wealthy family, and enjoyed the best education that France could give him. At the age of twenty-one he succeeded to a handsome property, at the very time when Linnæus, with an allowance of £8 a year from his father, was a struggling student at the University of Upsala, putting folded paper into the soles of his old shoes to keep out the damp and cold. Buffon was a man with a very keen interest in Natural History, and of remarkable industry and perseverance. In 1739 he was appointed Superintendent of the Jardin des Plantes at Paris, a post which he held till his death. In his great book on Natural History he gave a comprehensive account of all that was known concerning the distribution, habits, instincts, and structure of animals. Buffon strongly disapproved of the sharp lines and rigid systems of Linnæus, and was more fond of general theories than of minute details. His materials were derived rather from extensive reading than from direct observation, and, having weak eyesight himself, most of his anatomical work was done by assistants. His great reputation was due very largely to the exceedingly attractive style in which he wrote, and the charm with which he invested the whole subject. He led many to think about and take an interest in Natural History, and to add to it by their own observations, who would not otherwise have done so. Buffon must not, however, be thought of as a mere popular writer ; he really opened up new fields and led the way in many matters of the utmost importance. For instance, concerning the homologies of the mammalian skeleton, he was the first to compare the arm of man with the fore-leg of the horse. He also paid much attention to geographical distribution, and laid stress on the resemblance between the fauna of Northern Europe and that of America, and explained this by the existence of a former land connection wide enough to permit migrations. More-

over, the natural history of the various races of man
was for the first time treated scientifically. Buffon,
while at first a believer in the absolute fixity of
species, later on was led to suggest that plants and
animals may not be bound by fixed and immovable
limits of species, but may vary freely, so that one
kind may gradually and slowly be evolved by natural
causes from the type of another. He points out the
fundamental likenesses of type in many animals,
underlying the external diversities of character and
shape, which strongly suggest the notion of descent
from some common ancestor. He goes further than
this, and, granting the possibility of modification, sees
no reason to fix its limits. He even suggests that
from a single primordial being Nature has been able
in the course of time to develop the whole continuous series of existing animal and vegetable life.
This view is often expressed in a studiously guarded
manner, and denied in half-ironical terms a few pages
further on. This was the starting-point of the great
idea of Evolution, and Buffon's name well deserves a
place in its history. Before pursuing the progress of
the theory of Evolution we must consider the work
and personality of one of the greatest of all zoologists,
Cuvier, and of his contemporaries.

Cuvier, 1769–1832, was a man of extraordinary
industry and ability, and of commanding power—one
of the giants of biological science. He was born in
France, of Swiss parentage, and originally intended
for the Church. From 1788 to 1794 he was engaged
as private tutor to a French family near Caen. At
this time he commenced to study the fossil bra-

chiopods, which led him to compare them with the living species. Having by this time attracted the notice of influential people, he was in 1794 invited to Paris, and was appointed Assistant Professor, and in 1802 Professor, of Comparative Anatomy at the Jardin des Plantes. After this he rapidly rose to posts of great importance and distinction. Cuvier was specially distinguished by the zeal with which he applied himself to the actual dissection of large numbers of animals of many groups, by the clearness with which he kept himself free from the vague theories of his day ; also by the way in which he kept to his facts, drew his own conclusions from them, and absolutely rejected any theories that were opposed to them.

Cuvier may justly be regarded as the father of Comparative Anatomy. His most important service was the demonstration of the true nature of fossils. About the time of his arrival in Paris attention was being directed to the skeletons and parts of skeletons of animals which were being disinterred round about Paris, especially at Montmartre. These constituted a great puzzle at the time, the bones of many of them being immensely large and unlike those of any known animals. Cuvier eagerly set to work at this subject, and the minute knowledge he had obtained of living animals rendered the work comparatively easy, while the law of correlation—a single bone giving the clue to the structure, position, habits, food, &c., of the animal—helped him greatly. Cuvier was soon able to point out that the animals of which these were the remains were in many cases not like those now living in Europe; in other cases they were unlike animals living anywhere on the earth at the present time. Some, such as the elephant, rhinoceros, and opossum were no longer European ; others, such as the mammoth, were animals which at the present time are extinct. Attention was thus directed to this subject, and inquiry in other directions stimulated. In the older rocks other fossil remains still more unlike existing forms were discovered.

Cuvier has exercised the greatest influence on the study of zoology down to the time of Darwin. He was the founder of Comparative Anatomy, and penetrated into the subject much more deeply than Linnæus, his object being not merely to systematise but to study the animals themselves.

The real nature of fossils was known to Aristotle, but the knowledge was forgotten and lost. The current doctrine of the Middle Ages, and even so late as the eighteenth century, was that they were freaks of Nature, and they were regarded as unsuccessful creative attempts or models into which life had never been breathed. Cuvier overthrew all this finally, and proved that they were the remains of animals formerly dwelling on earth, and of different kinds or species to those now living. He, moreover, showed that the farther back we go in time, the more do they differ from recent animals. This was a tremendous step forwards, yet he stopped short, and held back on the brink of a great and comprehensive theory. He was struck with the differences rather than the resemblances among animals, and strenuously denied the possibility of any relation between recent and fossil forms, stating that "the immutability of species is a necessary consequence of the existence of scientific natural history." He formulated the doctrine of *Catastrophism*, or the periodical annihilation of existing animals followed by re-creation in a modified form. The question of the Origin of Species was now becoming a burning one.

GOETHE, 1790, in an "Essay on the Metamorphosis of Plants" showed the principle of fundamental unity, and demonstrated that all parts of a flower are really modified leaves or stem. In the cultivated rose the stamens and pistils are turned into petals, and gardeners find it possible to cultivate a plant so that it shall be all leaves and no flower, or shall have a gorgeous flower while the leaves remain small and insignificant. It is a pleasant reflection to a naturalist that the keenest and brightest intellects of all ages have not been unmindful of the charms of Natural History, and that they have taken delight in, and have themselves made contributions of great value to the subject. It is grateful to acknowledge this indebtedness to men more widely known for their labours in other directions. Aristotle's first classification of animals on scientific principles—viz., into Vertebrates and Invertebrates—holds good to the present day ; he also recognised the true nature of fossils as the remains of formerly living animals. To poets of all ages and every nation we owe many shrewd and accurate observations, especially on the habits of birds and flowers. Goethe was much struck with the power of modification or adaptation, which led him to the conception of blood relationship between animals, and of the descent from a common original type.

The most famous of the opponents of Cuvier, and upholders of the doctrine of mutability of species, were two of his own countrymen, who were colleagues in Paris for many years, Lamarck and St. Hilaire.

LAMARCK, 1744–1829, was originally intended for the Church, like Linnæus, Cuvier, St. Hilaire, and Darwin. He, however, had a passion for the army, and on the death of his father in 1760 set off for Germany, where the French were then fighting. There he distinguished himself as a volunteer, but owing to being accidentally disabled by a comrade had to abandon his career. He then went to Paris, and commenced the study of medicine, while holding a post as a banker's clerk. He was for a long time interested in botany, and in 1779 published three small volumes on the Flora of France. This attracted the notice of Buffon, and in 1793 Lamarck obtained an appointment at the Jardin des Plantes, the year before Cuvier, and applied himself with great vigour to the study of zoology. Lamarck and Cuvier worked practically side by side for many years. Cuvier, working mainly at Vertebrates and at fossils, was impressed by the differences between species, especially those between recent and fossil species. Lamarck worked mainly at the lower Invertebrates—jelly-fish, worms, and snails. He, on the other hand, was struck by their resemblances rather than the differences between them, and was impressed by the difficulty of settling which were distinct species and which might have come from the same parents. On this point he observes : " The more we know of animals and plants, the more difficult we find it to settle which are related to each other and which are not."

Lamarck was also impressed by the variability of animals and plants, according to their surroundings, and by the influence of drier soil or mountain habitat in causing stunting of growth and other alterations. In his " Philosophie Zoologique," published in 1809, and in the " Histoire Naturelle des Animaux sans Vertébres," published in 1815, he upholds the doctrine that all species of animals, including man, are descended from other species. With respect to the manner of modification, he attributes something to the direct action of environment, and much to use and disuse—*i.e.*, to the effects of habit ; for example, the use of the long neck of the giraffe for browsing on trees. He writes : " The systematic divisions of classes, orders, families, genera, and species are the arbitrary and artificial productions of man. Species arise out of varieties. In the first beginning only the very simplest and lowest animals and plants came into existence ; those of a more complex organisation only at a later period. The course of the earth's development and that of its organic inhabitants was continuous, and not interrupted by violent revolutions. Life is purely a physical phenomenon." If man can make such changes in a few hundred years, as for example, to produce the various domestic races of pigeons or rabbits, "is it not possible that nature in all the long ages during which the world has existed may have produced the different kinds of plants and animals by gradually enlarging one part and diminishing another, to meet the wants of each ? "

This is a full statement of all the essential points ; the unity of active causes in organic and inorganic nature ; the ultimate explanation of these causes in the chemical and physical properties of matter ; the derivation of all organisms from some few most simple forms ; the coherent course of events in Nature, and the absence of cataclysmal revolutions. It is a full and complete statement of the doctrine of Evolution as held at the present day, man himself being included, both as regards his mental powers and his bodily structure. This is often confused with the Darwinian theory, but is really quite distinct. Lamarck tells us that animals are descended one from another, and have a common bond of union or blood relationship. This explains the affinities of animals and of man, but fails to explain how and why. The Darwinian theory explains how this came about, and why this progressive transformation of organic forms has taken place, and what causes effected the uninterrupted production of new forms. Lamarck's views, though perfectly correct, were mere speculations until Darwin supplied the reason and explained the mode of action. Lamarck considered the long neck of the giraffe as due to its constantly stretching its neck to pick leaves from high trees ; the long tongue of the woodpecker, humming-bird, and ant-eater to the

habit of fetching food out of deep or narrow
crevices; the webbed toes of the frog to its constant endeavours to swim, and to the very movements of swimming. The true explanation, as we
shall see afterwards, was furnished by Darwin's
theory of natural selection.

ST. HILAIRE, 1771–1840, was educated as a priest,
but owing to his passionate love for zoology was
allowed to stay in Paris and work at the Jardin des
Plantes. He was offered an appointment, and
afterwards joined Lamarck at the Musée d'Histoire
Naturelle in 1793. St. Hilaire was a great friend
of Lamarck, and adopted his theory of descent.
He believed that the transformations of animals
were effected less by the action of the organism
itself, than by change in the outer environment.
A long and fierce controversy raged between
the three friends, Cuvier, Lamarck, and St. Hilaire,
chiefly between Cuvier and St. Hilaire, who,
strangely enough, was thought an abler man than
Lamarck. Shortly after Lamarck's death a formal
and final discussion took place in the Academy
of Sciences at Paris between Cuvier and St. Hilaire.
This occurred on the 22nd of February 1830,
and was renewed on the 19th of July, a bare week
before the outbreak of the French Revolution.
All Europe was excited by the controversy, and
none more so than Goethe, then an old man, and
a firm believer in the doctrine of Evolution.

Cuvier was far too strong for his opponent; a
hard hitter, and a man of greater personal power
and influence, and one who did not scruple to use

his full strength. He urged the evidence of mummies and other buried remains, which, after a lapse of thousands of years, agree in the smallest details with existing species. If a changing environment causes alterations, why, he asked, are these not altered ? He demanded evidence of connecting links between fossils and recent forms, and quoted his own unrivalled experiences as to its absence. Cuvier crushed his opponent by superior knowledge, by the better management of his case, and by personal authority. The verdict was a definite one, and the controversy was regarded as closed by final decision. The fixity of species was regarded as proved, and France has hardly yet recovered from the traditions of Cuvier.

We now approach the final stage in the great controversy, and the scene of action shifts to our own country. Light was first afforded, not by zoology or botany, but by the sister science of geology, a peculiarly British study, and a very recent addition to the tree of knowledge.

HUTTON, 1726–1797, when sixty-two years old, published his "Theory of the Earth." The main motive of this book was to show that in order to understand how the earth's crust, with its component layers, was formed, and how fossils got into them, we must not guess, but must look for ourselves, and see what is now going on around us—how rivers and glaciers are carrying down earth and stones from the mountains to the sea, how the solid earth is being wasted every day, and new rocks formed by the disintegration of older ones.

CHARLES LYELL, 1797–1875, like Hutton, was a Scotchman, and was born in the year of Hutton's death. Lyell worked out in detail Hutton's suggestions, and collected with great care all that was known of the changes now going on in the world, and of the causes that produce them; the rate of denudation and the amount carried down by streams; the mode in which plants and animals are buried in mud, peat, and sand. The first volume of his " Principles of Geology " was published in January 1830, seven months before the Paris controversy. In it he dealt with the changes continually taking place on the earth's surface ; the rising up and the subsidence of the earth's crust ; the action of rivers and volcanoes ; and showed how the present configuration of the earth was due to these causes. He pointed out that causes now in action and influences now at work are not merely competent to produce the present state of things, but must inevitably have done so. He showed the history of the earth to be continuous and uninterrupted, and that to explain its present condition and past history we have simply to look around us.

Lyell's facts were numerous and his reasoning cogent; his conclusions therefore steadily gained supporters. It is a curious fact that, as regards fossils, Lyell himself declined to apply to them the principles he so justly insisted on for the crust of the earth. Yet it was precisely by applying a similar train of reasoning to the further problem that the final solution was attained, and by the study of what is going on around us at the present day, that principles were determined competent to account for the changes in all past time, and the death-blow given to Cuvier's views.

Evidence as to the reality of Evolution was now rapidly accumulating, not only through the work and writings of those already mentioned, but from many other sides. Professor HUXLEY in 1859 refers to the hypothesis that species living at any time are the result of the gradual modification of pre-existing species, as the "only one to which physiology lends any countenance." Sir JOSEPH HOOKER in 1859, in his "Introduction to the Australian Flora," admits the truth of descent and modification of species, and supports the doctrine by many original observations.

HERBERT SPENCER's essay in the *Leader*, 1852, constitutes "the high-water mark" of Evolution prior to Darwin : "Even could the supporters of the development hypothesis merely show that the pro-duction of species by the process of modification is conceivable, they would be in a better position than their opponents.   But they can do much more than this ; they can show that the process of modification has effected and is effecting great changes in all organisms subject to modifying influences. . . . . They can show that any existing species—animal or vegetable—when placed under conditions different from its previous ones, immediately begins to under-go certain changes of structure fitting it for the new conditions. They can show that in successive generations these changes continue until ultimately the new conditions become the natural ones.   They can show that in cultivated plants and domesticated animals, and in the several races of men, these changes have uniformly taken place. They can show that the degrees of difference so produced are often, as in dogs, greater than those on which distinctions of species are in other cases founded. They can show that it is a matter of dispute whether some of these modified forms *are* varieties or modified species. And thus they can show that throughout all organic Nature there is at work a modifying influence of the kind they assign as the cause of these specific differences; an influence which, though slow in its action, does in time, if the circumstances demand it, produce marked changes; an influence which, to all appearance, would produce in the millions of years, and under the great varieties of condition which geological records imply, any amount of change."

It is impossible to depict better than this the condition prior to Darwin. In this essay there is full recognition of the fact of transition, and of its being due to natural influences or causes, acting now and at all times. Yet it remained comparatively unnoticed, because Spencer, like his contemporaries and predecessors, while advocating Evolution, was unable to state explicitly what these causes were.

We have now traced the main steps in the history of the doctrine of evolution, and have mentioned the names of the chief men with whom this history is most closely associated. This doctrine was rendered possible by Linnæus by the introduction of definite and precise nomenclature in the language common to all civilised nations, which thereby enabled men to speak of animals and groups of animals with exactness and certainty.   The difficulties of framing definitions based on facts, of which equally competent men took widely different views, led to the consideration of the question—Are species fixed or mutable? Buffon, Goethe, Lamarck, St. Hilaire, and Herbert Spencer are perhaps the most famous names among the supporters of Evolution. Lamarck and St. Hilaire are specially noteworthy, as they recognised the necessity of explaining the causes of the modification, and attempted, though with very partial success, to supply the explanation. Among the opponents of Evolution, Cuvier was by far the ablest.

The question first became a prominent one about the commencement of the present century. It was hinted at earlier by Buffon, but first obtained definite expression from Lamarck in 1801, and in more detail in the "Philosophie Zoologique," published in 1809.

It is very commonly assumed that the doctrine that animals are not immutable, or the doctrine of Evolution, is of very recent origin, and that for it we are indebted to Darwin. Nothing can be more erroneous, for, as we have seen above, not only was it very clearly and emphatically maintained by several writers at the commencement of the present century, or the conclusion of the last, but the idea is found stated more or less explicitly by Aristotle over 2000 years ago.

The "Doctrine of Evolution" teaches that there is a relationship between the animals of successive

periods or ages of exactly the same kind as that
which exists between the men of successive generations or centuries—viz., a blood relationship. Just
as men of each century are descendants of those of
a preceding century, and progenitors of those of later
ones, so it is with animals throughout all geologic
history.

It is well to point out clearly the difficulty which
has to be met. Animals of successive ages are
unlike, and fossils do not give the intermediate
series, nor satisfactory indications of them. The
problem we have to consider is this: The men of
successive generations are unlike in language, customs, dress, and appearance; now, are the differences
between animals of successive ages of the same
character as between men, though of wider nature;
or are they of such a kind as to forbid the idea of
descent one from another? In other words, are
species immutable or variable? The doctrine of
Evolution requires that they should be variable. In
order to establish this doctrine it must be shown
that there are causes, actually existent causes, competent to give rise to modifications of animals such
as we find in passing from one geologic age to
another. This is what is effected by the "Darwinian Theory," or the "Theory of Natural Selection," propounded independently and simultaneously
on July 1, 1858, by Darwin and Wallace.

CHARLES DARWIN was born in 1809, and studied
at Cambridge from 1827 to 1831. The voyage of
the *Beagle* occupied from 1831 to 1836, the greater
part of the time being spent on the east and west

coasts of South America, and the voyage round the
world being completed by way of New Zealand,
Australia, Mauritius, St. Helena, and Brazil, and
then by the Cape Verde Islands to England. The
remainder of his life was devoted to work in
England, work of extraordinary amount and most
varied character. By this work he slowly accumulated facts, and especially the conditions under which
the breeds of domesticated animals and cultivated
plants come into existence, and are propagated or
modified. I propose to consider more in detail the
history of Darwin's life in the last lecture.

ALFRED RUSSEL WALLACE was born in Mon-
mouthshire in 1823. As a boy he was an eager
naturalist. From 1844 to 1845 he was English
master at the Collegiate School at Leicester, and
while there made the acquaintance of Mr. H. W.
Bates, an ardent entomologist. A few years later
the desire to visit tropical countries became too
strong to resist, and a joint expedition took place
to collect Natural History objects. and to "gather
facts towards solving the problem of the Origin of
Species." In 1848 he started to the mouth of the
Amazon, and worked with Bates till 1850, when
Wallace moved to Rio Negro, finally returning in
October 1852. His vessel was destroyed by fire,
and he spent ten days in an open boat on the sea
in the mid-Atlantic. From 1854 to 1862 he spent
his time in the Malay Archipelago, where animal
life was most luxuriant and least affected by man.
In June 1858 Darwin received from Wallace the
MS. of a paper "On the Tendency of Varieties

to depart indefinitely from the Original Type," this being the same conclusion at which Darwin himself had arrived. Darwin wished to publish Wallace's paper at once, but on the urgent persuasion of Sir Joseph Hooker and Sir Charles Lyell he consented that extracts from one of his earlier MS. prepared in 1844, and from a letter to Professor Asa Gray in 1857, should be read at the same time. This took place at the Linnæan Society on July 1, 1858. Each man's discovery was perfectly independent, and neither knew on what lines the other was working. Full mutual recognition ensued, and most cordial intercourse and esteem.

SUMMARY.—We have now traced the gradually increasing tendency towards a belief in Evolution. This was suggested in a tentative and almost cynical way by Buffon, and warmly supported by Goethe, and during the first half of the present century by Lamarck, St. Hilaire, Herbert Spencer, and others. Owing to one fatal flaw this belief failed to command anything like general acceptance. We know from fossils that the former dwellers on earth were unlike those now living, and that the doctrine of Evolution therefore involves modification. No one was able to point to the causes which could lead to such modification, or to explain how it could have come about. This was the objection which was driven home with relentless force and persistency by Cuvier, supported by all the weight of his personal authority, and the influence rightly gained by his splendid contributions to the science of Comparative Anatomy. This is the objection

which in 1830 proved fatal, and which led to the
triumph of Catastrophism over Evolution. The supporters of Evolution were silenced, but not convinced.
In France the defeat was complete, but not so in
other countries. It was clear that the attack must
be made along new lines if there were to be any
prospect of success. This fatal objection must be
met. Was it not possible to determine the causes?
and if so, how? What could we hope to know of
causes which could lead to modifications in animals
of former geologic ages? Yet the answer was at
hand, for seven months before Cuvier's final triumph
at the Academy of Paris on July 19, 1830, appeared the first volume of Lyell's "Principles of
Geology," in which the true path was indicated, and
the key to the past shown to be afforded by the
study of the present. If we would know what
happened in former times, we should look around us
and see what is taking place before our eyes. In
this way Lyell gauged the forces of Nature—the
power of running water, the force of the tides, the
effect of frost and heat, the slow movements of
upheaval and subsidence, the slow change of climate
due to astronomical causes. He was able to prove
that causes now acting, and causes which must have
been in action from immeasurably remote periods,
were competent to produce the effects we wonder
at—the upheaval of mountain ranges, the excavation
of valleys, &c.—without any need for external or
supernatural agencies, and indeed, leaving no room
for such agencies, for then there was nothing further
to accomplish. The final stroke was given by

Darwin and Wallace, who, working perfectly independently, set themselves deliberately to attack the problem on the lines laid down by Lyell—viz., by prolonged and detailed study of the conditions under which animal life exists at the present day. After years of patient work they were led independently to identical conclusions, which were announced simultaneously from opposite sides of the globe.

POUTER CARRIER BARB FANTAIL TUMBLER ROCK PICEON TRUMPETER JACOBIN TURBIT

Fig. 1

LECTURE II

I NOW propose to consider the law to which Darwin and Wallace were led, the evidence upon which it is founded, and the conclusions which follow from it. In the method of attack I propose to follow Darwin, and I would warn you against almost inevitable disappointment, for it is with common-place things and facts of every-day occurrence that a great theory has to deal.

ARTIFICIAL SELECTION.

DOMESTIC PIGEONS.—Darwin early in his inquiry felt the importance of having individual animals under close observation, so that all conditions influencing them could be determined. For this purpose domestic animals were far more suitable than wild ones, and pigeons were selected for special study for these reasons :—(1) The evidence of their descent from a common ancestor is clear ; (2) Their historical records extend back many centuries; (3) Their variations are very great, all kinds being easily kept in captivity and all breeding true.

There are probably at least 200 kinds of pigeons

known which breed true, and these differ constantly from each other. The chief varieties are the following. (See Fig. 1.)

The *Pouter* is a large and upright bird with a long body and long legs, a moderate-sized beak, and a very large crop and œsophagus. It has the habit of inflating its crop, producing a "truly astonishing appearance," being then "puffed up with wind and pride."

The *Carrier* is a large bird with a very long beak. The skin round the eyes, over the nostrils, and on the lower jaw is much swollen, forming a prominent wattle.

The *Barb* has a short and broad beak, and a wattle of moderate size.

The *Fantail* has tail feathers to the number of 34 or even 42, twelve being the normal number. The tail is expanded and held erect. It has a peculiar gait, and a curious habit of trembling by convulsive movements of the neck. In a good specimen the tail should be long enough to touch the head.

The *Turbit* has a frill formed by divergent feathers along the front of the neck and breast. The beak is very short.

The *Tumbler* has a small body and short beak. During flight it has the habit of turning involuntary back somersaults.

The *Jacobin* has long wings and tail and a moderately short beak. It has a hood formed by the feathers of the neck.

The *Trumpeter* has a tuft of feathers at the base of the beak, curling forwards. The feet are much feathered. The coo is very peculiar, and unlike that of any other pigeon, being rapidly repeated and continued for several minutes.

Among these forms there is thus great diversity in both form and colour. This diversity also affects the internal structure, for example the skull : the caudal and sacral vertebræ and also the ribs vary in number. The number of primary wing and tail feathers, the shape and size of the eggs, the manner of flight, and almost all other characters, also differ. If these birds were now found in a wild state, they would be considered to constitute distinct genera, yet they are known to be all descended from *Columba livia*, the blue rock-pigeon of Europe, Africa, India, &c.

The arguments brought forward by Darwin to prove this are as follows :—

(i.) All domestic races are highly social, and none of them habitually build or roost in trees ; hence it is in the highest degree probable that their ancestor was a social bird nesting on rocks.

(ii.) Only five or six wild species have these habits, and nearly all these but *Columba livia* can be ruled out at once.

(iii.) *Columba livia* has a vast range of distribution—from Norway to the Mediterranean, from Madeira to Abyssinia, and from India to Japan. It is very variable in plumage and very easily tamed. It is identical with the ordinary dove-cot pigeon, and except in colour practically identical with toy pigeons generally.

(iv.) There is no trace of domestic pigeons in the feral condition.

(v.) All races of domestic pigeons are perfectly fertile when crossed, and their mongrel offspring are also fertile. Hybrids between even closely allied *species* of pigeons are, on the other hand, sterile.

(vi.) All domestic pigeons have a remarkable tendency to *revert* in minute details of colouring to the blue rock-pigeon. This is of a slate-blue colour, with two bars on the wings, and a black bar near the end of the tail. The outer webs of the outer tail-feathers are edged with white : these markings are not seen together in any other species of the family. This tendency to *revert* was demonstrated by Darwin as follows : He first crossed a white fantail with a black barb ; then a black barb with a red spot (a white bird with a red tail and a red spot on the forehead). He then succeeded in crossing the mongrel barb-fantail with the mongrel barb-spot, and the birds produced were blue, with markings on the tail and wings *exactly like those of the ancestral rock-pigeon.* Thus two black barbs, a red spot, and a white fantail, produced as grandchildren birds having every characteristic of *Columba livia*, including markings found in no other wild pigeon.

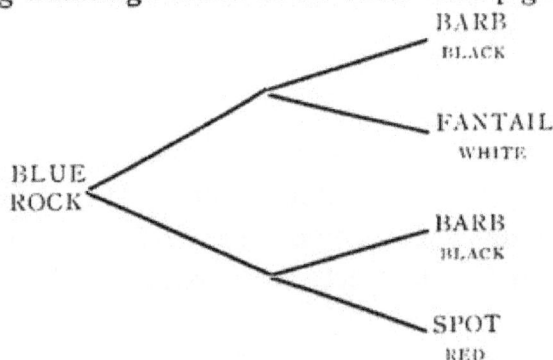

DORKINC

COCHIN

HAMBURC

CAMECOCK

BANTAM

POLISH

SPANISH

SILK

Fig. 2

(vii.) All domestic pigeons resemble *Columba livia* in their habits. They all lay two eggs, and require the same time for hatching. They prefer the same food, and coo in the same peculiar manner, unlike other wild pigeons.

(viii.) *Columba livia* has been proved to be capable of domestication in Europe and in India.

(ix.) *Historical Evidence.*—Referring to Aldrovandi, who figured pigeons in the year 1600, we find the Jacobin with a less perfect hood; the Turbit apparently without its frill ; the Pouter with shorter legs, and a less remarkable bird in all respects ; the Fantail with fewer tail feathers, and a far less singular appearance ; the Tumbler existed then, but none of the short-faced forms ; the Carrier had a beak and wattle far less developed than the modern English Carrier. These were the same groups of pigeons, but with their distinctive characters less marked, thus showing convergence towards their common ancestor.

The mode of action of these changes is by *artificial selection*, or the power possessed by man of influencing the shape, size, and colour of animals by the accumulation of small differences in successive generations. This depends on two laws :

1. *The Law of Variation*, depending on the fact that no two animals are exactly alike.

2. *The Law of Inheritance*, or the tendency to hand down characters and peculiarities to descendants.

DOMESTIC POULTRY.—These afford another good example of artificial selection. (See Fig. 2.)

The *Gamecock* is characterised by its upright comb, strong beak, sharp spurs and by its great

FIG. 3.

Gallus bankiva (The Jungle-Fowl).

courage. Of all the different forms, this most resembles the wild *Gallus bankiva*.

The *Cochin* is of large size, and scarcely able to

fly. The plumage is soft and downy; the legs thick
and feathered; the comb and wattle well developed.

The *Dorking* is a large bird with a large comb
and wattle, and possesses an extra toe.

The *Spanish* is tall and of stately carriage. The
comb and wattle are very large.

The *Hamburgh* has a flat comb prolonged backwards, and a moderate-sized wattle.

The *Polish* is characterised by a large crest of
feathers, the comb being either absent or very small.
The wattle is sometimes replaced by a tuft of
feathers.

The *Bantam* is of small size and bold erect
carriage.

The *Silk-fowl* is a small bird with very silky
feathers.

All these birds, differing so much among themselves, are descended from *Gallus bankiva*, the
Jungle-fowl (Fig. 3), which is still found in a wild
state in India and the Malay Islands. This bird
was domesticated in India and China before 1400 B.C.,
and was introduced into Europe about 600 B.C.
Several distinct breeds were known to the Romans
about the commencement of the Christian era.

THE ANCON SHEEP.—Another well-known example of artificial selection, and one of the few
known instances in which new breeds have suddenly
originated, is that of the *Ancon* Sheep, bred by Seth
Wright, a farmer of Massachusetts. In 1791 one of
Seth Wright's sheep bore a male lamb which had
very short and bandy legs. Now, as Wright was
continually losing his sheep, owing to their jumping

over his fences, it occurred to him that, if he could produce a breed of sheep with short bandy legs, he would lose none of them, as they would be unable to jump his fences. He therefore bred entirely from the short-legged ram when it had reached maturity, and after a few years succeeded in raising a considerable flock of this variety, which was known as the Ancon sheep.

The power of artificial selection is almost unlimited, and breeders of animals speak with the utmost confidence of being able to produce any desired result in the form of the body ; and, in the case of poultry and pigeons, in the length of beak, the number of feathers, and even in the markings on particular feathers. Lord Spencer says : " It is therefore very desirable before any man commences to breed either cattle or sheep, that he should make up his mind as to the shape and qualities he wishes to obtain, and strictly pursue this object." And speaking of Leicester sheep Lord Somerville remarks : " It would seem as if they had just chalked on the wall a form perfect in itself, and then had given it existence." So also with plants ; enormous changes are effected by cultivation—*i.e.*, by selection, in fruits and flowers.

NATURAL SELECTION.

The theory of Natural Selection teaches that there are in Nature causes which act in much the same way as man acts when selecting artificially the best animals for breeding purposes ; causes which must lead to structural modifications; and that this is the clue to the unlikeness between the fauna of successive geologic ages.

I propose first to give an outline of the argument and to consider the arrangement given in Mr. Wallace's chart of the Theory of Natural Selection.

WALLACE'S CHART OF THE THEORY OF NATURAL SELECTION.

PROVED FACTS.	CONSEQUENCES.
A. RAPID INCREASE OF NUMBERS—	STRUGGLE
B. TOTAL NUMBERS STATIONARY—	FOR EXISTENCE
C. STRUGGLE FOR EXISTENCE —	SURVIVAL
D. VARIATION WITH HEREDITY —	OF THE FITTEST
E. SURVIVAL OF THE FITTEST —	STRUCTURAL
F. CHANGE OF ENVIRONMENT —	MODIFICATIONS.

A. *All animals produce far more young than can survive.* Consider for instance the enormous number of eggs of fish or oysters.

B. *The total numbers are on the average stationary.* As a necessary consequence of this there is a *struggle for existence*, because there is neither enough space nor food for all.

D. *Variation with heredity.* No two animals are exactly alike, and their distinctive characters are transmitted from generation to generation. The consequence of this is the *survival of the fittest*—that is, that in the long run those best adapted to their circumstances and environment will have the best chance of surviving, and of leaving descendants who

will hand down their peculiarities. Just as man
selects artificially the forms best suited for his purpose, and by breeding from them produces great
changes in structure and habit, so in Nature the best
and fittest of each generation have an advantage and
the best chance of survival.

F. *Change of environment*, rendering old characters of less value and bringing new ones to the fore.

From this follow—*Structural modifications.*

Causes are always at work which must lead to
change in structure, and this to an apparently unlimited extent.

Let us now examine the argument more closely.

A. RAPID INCREASE OF ORGANISMS.—"There is no
exception to the rule that every organic being, animal
or plant, naturally increases at so high a rate that,
if not destroyed, the earth would soon be covered by
the progeny of a single pair." Man himself has
doubled his numbers in the United States in the
course of twenty-five years, and at this rate in less
than 1000 years there literally would not be standing-
room on the earth for his progeny. Linnæus showed
that an annual plant producing two seeds only—and
there is no plant so unproductive as this—and these
each producing two in the following year, and so on,
would in twenty-one years produce over a million
plants, as shown in the following table :—

ANNUAL PLANT PRODUCING TWO
SEEDS ONLY.

YEAR.	NO. OF PLANTS.
1	1
2	2
3	4
5	16
7	64
9	256
11	1,024
13	4,096
15	16,384
17	65,536
19	262,144
21	1,048,576

The rate of increase of an animal, each pair producing ten pairs annually, and each animal living ten
years, is shown in the following table :—

YEAR.	PAIRS PRODUCED.	PAIRS ALIVE AT END OF YEAR.
1	10	11
2	110	121
3	1,210	1,331
4	13,310	14,641
5	146,410	161,051
10		25,937,424,600
20	Over 700,000,000,000,000,000,000	

Vast numbers of eggs are laid by some animals;
the conger-eel, for instance, lays 15 millions; the
herring 20,000; the oyster from half a million to 16
millions; and a very large oyster may produce even
60 millions of eggs. Supposing we start with one
oyster and let it produce 16 million eggs, the average
American yield, and let half, or eight millions, be
female and go on increasing at the same rate; in the

second generation we shall have 64 millions of
millions of female oysters. In the fifth generation—
i.e., the great-great-grandchildren of our first oyster—
we should have 33 thousand millions of millions of
millions of millions of millions of female oysters.
If we add the same number of males we should have
in all 66 + 33 noughts. If we estimate these as
oyster-shells, we should have a mass more than eight
times the size of the world.

A large number of eggs or young is, however, not
essential. The *Fulmar petrel* lays only one egg, yet
it is believed by Darwin to be one of the most
numerous birds in existence. The *Passenger* pigeon
again only lays two eggs, yet it is extraordinarily
abundant in parts of North America, where its enormous migrating flocks darken the air for hours. A
remarkable account is quoted by Wallace of a wood
in Kentucky, 40 miles in extent, where there was a
perpetual tumult of crowding and fluttering pigeons,
and where there were as many as a hundred nests
on a single tree, the branches of which were often
broken off by their weight, and the ground strewn
with broken limbs of trees, eggs, and young birds, on
which herds of hogs were fattening. Hawks, buz-
zards, and eagles were flying about in large numbers,
seizing the young birds at pleasure; and numerous
parties of men from all parts of the adjacent country
were camping with their families for several days,
felling trees to get the nests.

Another good example of rapid increase in numbers was seen in the rabbit pest of Australia in
1887. The common grey variety of wild rabbit

introduced into Victoria in 1860, became so prolific as to overrun the greater portion of the colony, and great sums of money were expended in endeavouring to exterminate it..

B. THE NUMBERS ARE STATIONARY AS A WHOLE.— Some forms increase while others diminish in numbers, and hence there is not actually this rapid increase of adult forms.   Enormous numbers are devoured as food in their early stages, the seeds of plants being eaten by birds, and the young of various animals by other animals.   Many animals again have their numbers kept down by parasites ; for instance, the caterpillar of the large garden white butterfly is peculiarly liable to attacks from the ichneumon fly, which lays its eggs in the body of the caterpillar ; and out of 533 larvæ collected by Mr. Poulton in 1888, 422 full-fed caterpillars died from the presence of ichneumon grubs—*i.e.*, four out of five perished from this cause alone.

C. THE STRUGGLE FOR EXISTENCE.—Every single organic being may be said to be striving to the utmost of its power to increase its numbers, while the vast majority of animals and plants that come into the world are doomed to die early.   For instance, in the case of an annual plant producing 1000 seeds —which is no very large estimate—if the numbers remain stationary, only one of these 1000 can on the average come to maturity ; and it may be said to struggle with plants of the same or other kinds which already clothe the ground.   In fact, "all the plants of a country are at war with each other."

Again, the introduction of goats into St. Helena led

to the entire destruction of the native forests, consisting of about a hundred distinct species of trees
and shrubs, the young plants being devoured by the
goats as fast as they grew up. A famous illustration
of the nice balancing of forces between animals and
plants is furnished by cats and clover. The red
clover is fertilised almost exclusively by humble
bees; and field mice destroy the nests of humble
bees in large numbers. Newman estimates that
two-thirds of the total number of humble bees'
nests in England are thus destroyed. Now, the
number of mice depends largely on the number of
cats, and hence the abundance of clover depends on
the proper supply of cats. Darwin remarks that
"battle within battle must be continually recurring
with varying success; and yet in the long run the
forces are so nicely balanced that the face of Nature
remains for a long time uniform, though assuredly
the merest trifle would give the victory to one
organic being over another."

The real struggle is between the most closely
allied, and therefore competing forms. The black
rat, for example, was the common rat of Europe till
the beginning of the eighteenth century, when it was
driven out by the larger brown rat. Competition is
keener in direct proportion to the closeness of
interests, the two covering the same ground. That
the struggle for existence is a very real one, and does
actually lead to the extermination of less fit forms, is
seen in the way in which the Maoris, for instance,
are gradually becoming exterminated. The struggle
is not necessarily one of actual warfare, the stronger

killing the weaker, although this may occur; and it is rather to more rapid multiplication and greater power of endurance that survival is due. So in the case of commercial competition among men, they do not actually fight; yet all cannot succeed, and failure means bankruptcy and starvation, and leads to destruction as surely as actual hand-to-hand warfare. Industrialism is in fact war under the forms of peace.

D. VARIATION.—It is a well-known fact that no two animals are absolutely alike; Darwin gives many instances of this. The Laplander by long practice knows by sight, and can actually name, each reindeer; the ants of one nest know and recognise one another; the sheep-dog picks out his own sheep unerringly; shepherds have won wagers by recognising each sheep in a flock of a hundred, which they had never seen till a fortnight previously. Voorhelm, an old Dutch florist, kept 1200 kinds of hyacinths, and was hardly ever deceived in recognising each variety by the bulb alone. The whole theory of breeding animals or of cultivating plants and fruits would fall to the ground, and selection would be impossible, unless variations occurred, and there were differences to select from.

Variability, which is a subject attracting much attention at present, is a general rule among animals, and applies to all of them. For example, Carpenter has shown that among the *Foraminifera* the range of variation includes not merely specific characters, but also generic and even ordinal ones. Among anemones great variations are found. Among snails

198 varieties of the common wood-snail have been
described. In the case of insects, many of our
common English butterflies vary enormously.
Among birds, a remarkable series of facts is
quoted by Wallace, from a memoir by Mr. Allen
on the birds of Florida. Exact measurements were
taken of large numbers of these, and all parts were
found to vary; not only were variations of 15 to
20 per cent in actual and relative sizes found
ordinarily; but the variations affected the length
and breadth of the tail and wings; the length,
width, depth, and curvature of the bill, the form of
the toes, the intensity of colour and nature of the
markings. This was therefore not a case of minute
or infinitesimal variations, but variations on a large
scale, affecting all parts and in all directions.

Variation in habits.—A good illustration of the
variation which may occur in habits is found in the
Kea, a curious parrot inhabiting the mountain ranges
of the island of New Zealand, and feeding naturally
on the honey of flowers, on insects, fruits and
berries, which till quite recently comprised its
whole diet. However, since the European occupa-
tion of the island, this bird has acquired a taste for
carnivorous diet with alarming results. It began by
picking the sheep-skins hung out to dry, or the meat
in process of being cured. In 1865 it was first
observed to attack living sheep, which were frequently found with raw and bleeding wounds in
their backs. Since then it is stated that the bird
actually burrows into the living sheep, eating its
way down to the kidneys, which form its special

delicacy. In consequence of this, the bird is being destroyed as rapidly as possible, and one of the rare and curious members of the New Zealand fauna will no doubt shortly cease to exist.

Variation occurs to a large extent also in plants, as seen by the different number of varieties which are described by different observers. Of the bramble, for instance, Bentham described five British species ; while Babington, about the same time, described as many as forty-five.

Man himself gives as good illustrations of variation as any animal; for instance, in his stature, habits, mental and bodily powers; in such individual details as minute inflections of the voice, or in the shape of the ear. To take the most recent development, Galton's work on finger-prints, the patterns formed by the ridges at the tips of the fingers and thumb are found to be unlike in any two cases, and to retain their peculiarities unchanged throughout life, thus forming one of the most trustworthy modes of identification yet discovered.

We read of the dead body of Jezebel being devoured by the dogs of Jezreel, "so that no man might say—This is Jezebel;" and that the dogs left only her skull, the palms of her hands, and soles of her feet. It is a curious satire that these parts should now be shown to be the very ones by which a corpse could be most surely identified.

The causes of variation are very imperfectly understood, and careful inquiries are now being made in order to elucidate them. Variation is undoubtedly influenced greatly by external conditions,

such as nutrition, cold, &c., and these may affect
the young, or even the embryo, or the egg before
it is laid by the parent. No two animals can
ever come into existence under absolutely identical
conditions ; neither can any two after birth be exposed to absolutely the same conditoins.

Variation under domestication is the rule instead
of the exception, and occurs more or less in every
direction. Consider, for instance, the extraordinary
variations in size and mode of growth of the
cabbage ; the solid heads of foliage utterly unlike
any plant in a state of Nature; the curiously wrinkled
leaves of the savoy, the purple leaves of the pickling
cabbage, the compact heads of flowers of the
broccoli and cauliflower, the curious stem of the
Kohlrabi, which grows like a turnip. Again, of the
apple there are at least a thousand varieties known,
all descended from the common crab-apple. In fact,
as Wallace says, "there is hardly an organ or a
quality in plants or animals which has not been
observed to vary ; and further, whenever any of
these variations have been useful to man, he has
been able to increase them to a marvellous extent
by the simple process of always preserving the best
varieties to breed from."

Limits to variation must of course exist, and it is
evident that up to some point or other variations
must be predetermined on definite lines. The
inconstancy of chemical composition or instability
is specially characteristic of living things. Variations are spoken of as accidental, not in the sense
of their not being all due to natural causes, but

inasmuch as they are accidental in relation to the sifting process of natural selection.

E. NATURAL SELECTION, OR THE SURVIVAL OF THE FITTEST.—Those animals which are most in harmony with their surroundings will survive.   Just as in the breeding of animals by artificial selection, those animals are selected to survive which have certain favourable peculiarities. usually too slight for any but a practised eye to detect; so under natural conditions the possession of some useful variation, such as a slight increase of speed, or power of endurance or strength, or a keener sense of vision, will determine which shall be the survivors in a large herd of animals.

The action of natural selection is well shown by the following example.   Many insects of Madeira have either lost their wings, or had them so much reduced as to be useless for flight, while their allies in Europe have them well developed.   The explanation of this is that Madeira, like other temperate oceanic islands, is much exposed to sudden gales, and the most fertile land being near the coast, the insects if able to fly are liable to be blown out to sea and lost.   Year after year the individuals which had the shortest wings, or which used them least, would have an advantage, and so would survive.   Hence the survival in the island of the insects with the smallest wings.

In Kerguelen Island, one of the stormiest places on the globe and a place entirely without shelter, all the insects are incapable of flight, and most of them are entirely destitute of wings.   These insects

include a moth, several flies, and many beetles. Now
these are the descendants of winged insects, which
must have reached the island by flying, and gradually
lost the power of flight, as in the insects of Madeira.

The importance of small variations.—We are apt
to overlook the importance that slight variations
may have, which is well shown in the artificial
breeding of animals. So it is with human affairs,
where important points, such as the fate of a Ministry,
or even the determination of peace or war between
two countries, often depends on side issues. In trade,
accidental variations may determine success by attracting attention. The success of a novel, play or
oratorio is often impossible to predict, and often
depends on a mere caprice. Change for the mere
sake of change may involve the misery or even
death of thousands, and cause alternating periods of
great prosperity and greater distress. This is well
seen in the changes of fashion in dress, which in the
case of the feathers for ladies' hats, or a particular
kind of fur, may mean destruction and wholesale
slaughter, even to extermination, of particular animals.

INHERITANCE.—The more favoured ones will not
only survive, but will tend to hand down to their
descendants their special advantages; and of these
descendants some will have these special peculiarities
in a less marked degree than their parents, some
equally and others more strongly marked. The
latter will in the long run survive, if the further
development of this special advantage confers further
benefit on the individual. The whole theory of the
breeding of animals depends on inheritance. For

instance, the pedigrees of race-horses are kept with most scrupulous care, and enormous prices are paid for horses for breeding purposes. So it is with pigs, poultry, dogs, and cattle, and in the improvements effected in fruits and flowers. Not only are good characters inherited, but bad ones also, and even diseases and malformations, such as insanity, gout, short-sight, cataract, and colour-blindness, among men.

F. THE INFLUENCE OF ENVIRONMENT.—We have seen that the struggle for existence results in the survival of the fittest. Now, if the conditions remain permanent there is no reason to suppose that the race would alter, for the fittest now would be so a thousand years hence, provided the external con-ditions did not change in the meantime. Variations would no doubt occur, but as none of these would confer an advantage, they would not be preserved. In a very few cases this is so, but constant change is the rule. For instance, consider the change effected in Australia by the arrival of civilised man with his dogs, horses, &c., resulting in the aboriginal inhabit-ants, human and animal alike, being killed off by competition. Man's influence is no doubt great ; but other influences are still more potent.

Here we derive much assistance from the evidence afforded by geology, which tells us that, as regards the earth we live in, things were not always as we find and know them now. The marks on boulders and deposits of glacial mud and clay, show that these boulders have been brought from afar, that their only possible means of transit was by glaciers, and

hence that our climate was once much colder than it
is at the present time. If, on the other hand, we
turn to Greenland, which is now in the glacial condition, we find beneath the ice, beds containing fossil
plants, showing the former existence in Greenland of
such plants as the chestnut, oak, plane, beech, and
poplar; nay even the magnolia, vine, walnut, and
plum; proving the former existence not only of a
moderate, but of a warm climate.

Geology shows us that the boundaries of land and
sea are not constant; for instance, that Britain and
France were once united, and that the sea is encroaching on the land on one side, while the land is
encroaching on the sea on the other. The crust of
the earth is made up chiefly of rocks deposited under
water; therefore where there is now dry land, there
must once have been open sea. Geology further shows
that these changes have not been of a sudden cataclysmal character, but gradual ones, changes which
are actually in progress at the present day, and
which must always have been going on since the
earth began.

The last link in the chain is now complete.
Owing to incessant geological change in environment, variations in structure, previously useless or
harmful, become advantageous, and their possessors
thereby triumph and survive, and hand down their
advantages to their descendants; thereby in course
of time causing structural modifications of greater or
less extent in the race. All Nature is in a condition
of more or less unstable equilibrium. The action of
environment is *indirect*, and changed conditions of

life bring to the front variations previously useless, a slight change often causing great results.

The fittest to survive is not necessarily the one most perfect ideally, but rather the one best adapted to, and most in harmony with, the environment at the time. To be too far ahead of the times is far more fatal, as regards worldly prospects in human society, than to be conspicuously behind them ; for in the latter case the individual is pitied and allowed the crumbs of charity ; in the former he is regarded with suspicion and starved. This may constitute a consoling thought to those who are temporarily out in the cold, and who see the place they covet occupied by manifestly inferior men.

COMPARISON OF NATURAL AND ARTIFICIAL SELECTION.—In comparing natural with artificial selection we meet with the same principles, and yet domestic races differ from natural ones. The reason of this is that man selects and propagates modifications solely for his own advantage or pleasure, and not for the creature's benefit ; he always tends to exaggerate, to go to the extreme point of selecting useful or pleasing qualities. Races of animals are thus produced which would be incapable of independent existence; for instance, the prize-pig, which has to be fed with a spoon like a baby, would have a poor chance of existence in the wilderness ; and races such as the Italian greyhound, the Fantail pigeon, hornless bulls, or the bull-dog of our dog shows, could not survive unless artificially protected.

Artificially-bred animals and plants are in fact in a condition of unstable equilibrium, and have a tendency to *revert* or slip back to a former and more stable condition, and are kept with difficulty at the stage they have reached. This is well seen in the tendency which crossed pigeons have to revert to the ancestral stage of the blue rock-pigeon. We may illustrate this point by a simple mechanical comparison.

A pack of cards lying on a table are in a condition of stability, and they may be taken to represent the normal or ancestral condition. If now the cards are built up to form a pagoda they are eminently unstable, although forming a more imposing structure, and are liable to collapse with the slightest touch, and revert to their former condition of stability. So the artificially-produced pigeons are much more imposing birds than the blue rock-pigeon, but they are in a condition of great instability, and readily revert to the ancestral condition.

Natural selection, on the other hand, acts, not for the good of man, but for the good of the species, and tends to preserve, develop, and perpetuate all characters which will give the species an advantage in the struggle for existence. There is no known instance of an animal or plant having either structure or instinct developed in order to benefit another species. Every species is for itself and for itself alone.

LECTURE III

In the first lecture we discussed the theory of Evolution, which claimed that animals now living are the descendants of those that lived formerly. We found the objection to the theory to be that animals which lived formerly were unlike those now living, and therefore that modification was necessary. In the second lecture, we saw how this objection was met by the theory of Natural Selection, and that causes were shown to be in existence not only competent to give rise to modification, but inevitably leading to it.

Let us now test this theory by seeing whether or not it is in accordance with the facts with which it has to deal. There is no possible doubt as to which series of facts we must deal with first. We must unearth these ancestors, put them in the witness-box, examine and cross-examine them, and see whether they support our case or not.

Let us first examine the Crust of the Earth, and its great division into Stratified and Igneous rocks. A headland, for example, consists of stratified rocks ; the waves eat away the shore, and the cliffs fall in ; streams carry down the mud and sand ultimately

into the sea, where it is deposited in a plane over
the sea-bottom. The nature of the deposit will
depend on the source of the supply. If we ask how
and why the cliff is stratified, the cliff itself will tell
us. The stratified condition is due to deposition
under water. Igneous rocks are intrusive and are
caused by the volcanic heat of the deeper parts of
the earth.

The crust of the earth is made up of sedimentary
or stratified rocks deposited one above another,
the most recently formed being at the top.
Their position may be subsequently disturbed, yet
the general relation can usually be determined.
Geologists find the sequence to be much the same,
and to show general agreement in all parts of the
earth, so that the same names can be employed.

Particular deposits may be thicker or thinner and
of variable nature in different localities, or absent
altogether. To interpret the crust of the earth
we must read it as a history of the earth in
successive chapters, like successive centuries or ages
in the history of mankind. The chief differences
are that it consists of several chapters of unequal
length, of which there is no means of determining
the absolute age or duration, separated by gaps
about which we have no record whatever. The
history of these times is revealed by fossils, "imprints
on the pages of time," which can be compared to the
descriptions and drawings in the written records of
man. Bones, teeth, shells, and other hard parts are
often found in extraordinarily perfect condition.
These records tell us, for example, that but a short

time ago, geologically speaking, the lion, bear, rhinoceros, mammoth, and hippopotamus lived in Britain.

This evidence is of great importance, for it consists of real remains of former inhabitants of the earth, who stand in the same relation to the present fauna as our Saxon or Norman ancestors do to ourselves; that is, they are the remains of the actual ancestors of living animals, and must include among them, were our collection complete, such ancestors of all living animals. This is evidence of peculiar value, and there is no gainsaying it. The true nature of fossils was neglected and greatly misunderstood until the early part of the present century; but since the time of Cuvier fossils have been collected diligently, and large numbers have been obtained from all parts of the earth. It is now known that the fossil Mollusca are considerably more numerous than those now living on the earth, and probably fossil mammals are almost as numerous as recent forms. The age of fossils cannot be determined absolutely; their relative age is, however, known, and we are able to draw up tables giving the order and sequence of events, though not the actual dates; sequences, moreover, that will apply not merely to one, but, with certain reservations, to all parts of the globe.

In order to learn the lesson taught by fossils let us take those found in England at different periods.

THE CRUST OF THE EARTH.

		FEET
TERTIARY OR KAINOZOIC: 3,100 ft.	Historic	
	Pre-historic	
	Pleistocene .	50
	Pliocene .	. 130
	Miocene .	. 440
	Eocene .	. 300
		2,180
SECONDARY OR MESOZOIC: 9,350 ft.	Cretaceous .	1,360
	Jurassic .	2,340
	Liassic .	900
	Triassic .	4,750
PRIMARY OR PALÆOZOIC: 111,750 ft.	Permian .	550
	Carboniferous	13,000
	Devonian .	16,200
	Silurian .	32,000
	Cambrian .	30,000
	Archaian .	20,000

Table of geological strata, with approximate thickness of the several strata in Britain. Total thickness about 23½ miles.

(1) THE TERTIARY OR KAINOZOIC PERIOD.—The pre-historic cave deposits found in the caves used as dens by wild beasts, who dragged into them the carcases of their prey, prove the occurrence of the *Hyæna, Cave-bear, Rhinoceros, Lion, Reindeer, Bison, Hippopotamus*, and *Beaver*—with man. These forms are now living, but not in England. In the Glacial period the earlier deposits show some of these disappearing, others persisting, and animals appearing that now do not exist anywhere—viz., the *Mammoth, Woolly Rhinoceros*, the *Sabre-toothed Tiger* and the *Irish Elk* or deer. In the Pliocene period we find evidence of the *Mastodon* and *Tapir;* and in the Eocene period, of the *Didelphys, Opossum*, and *Hyracotherium.*

(2) THE SECONDARY OR MESOZOIC PERIOD.—The entire group found at this period is extinct. The

Ammonites are characteristic secondary forms, and are very abundant, but none survive into the Tertiary period. The *Ichthyosaurus*, *Plesiosaurus*, and *Mosasaurus* are extremely characteristic, and confined to the Secondary period.

(3) THE PRIMARY OR PALÆOZOIC PERIOD.—Here we find fossil plants in the Coal measures, the *Lepido-dendron, Calamites, Sigillaria*, and *Stigmaria*. Fish are also found abundantly. The *Trilobites* are exclusively Palæozoic; also the Sea-scorpions, *Pterygotus*, and *Eurypterus*.

The general conclusions we arrive at are :

(1) There is a general advance in organisation from the lower to the higher or more recent deposits, and an increase in the diversity of type.

(2) There is no evidence of sudden breaks or cataclysms; there is no break between the Tertiary period and the present day. Some species die out and others appear, and some persist unchanged. The very evidence which Cuvier relied on to prove Catastrophism disproves it when examined more carefully and with fuller knowledge.

(3) Some forms, known as *persistent types*, remain unchanged for great periods. This constitutes no real difficulty, for Natural Selection does not of necessity involve progression or change of any kind, and is quite consistent with a stationary condition, provided that the environment, or at least all the features of the environment affecting them, remain unchanged. These are examples of the real aristocracy of animals, for they can date back their descent not merely to the time of the appearance

of man, but almost to the first appearance of animal life of which we have positive evidence.

As examples of *persistent types* may be cited *Globigerina*, which shares in the formation of chalk, and is found in the Trias, or bottom of the Secondary strata ; *Limulus*, the king-crab, which occurs in the Trias, and is found on the American coast at the

FIG. 4.   FIG. 5.

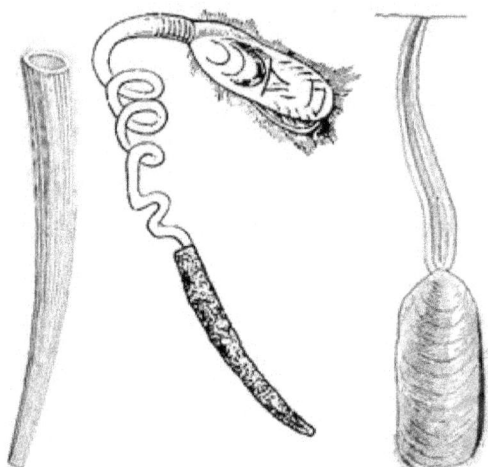

Dentalium.   *Lingula.*

present day ; *Dentalium*, the tusk-shell, an animal in some respects between an oyster and a snail, with a tubular conical shell about two inches long, by which it burrows in the sand, is found in the Devonian and perhaps Silurian strata ; the *Pearly Nautilus*, a rare animal, is found in the lower Silurian and at the present day ; *Lingula*, an animal with extreme tenacity of life, is found in the lower Cambrian and at the present day, and, so far as we can see, is unchanged.   (Figs. 4 and 5.)

Human customs and people have often been referred to for examples of the laws of biology, and there is no need to look beyond them. As an example of a persistent type, cannot we at once call to mind a nation, a homeless nation, the members of which occur in all countries, yet have no country of their own; a nation which, in spite of persecution of unexampled severity, endured not once only, but many times repeated and in most diverse forms, has held its own; a nation which, in spite of various and shifting environments to which it has been exposed, has with singular tenacity retained and preserved its language, traditions, and ceremonial observances in all essential features intact.

THE IMPERFECTION OF THE GEOLOGICAL RECORD.

Undoubtedly the most interesting and important point concerning fossils yet remains to be considered. Fossils are the former inhabitants of the globe, and therefore, on the Theory of Evolution, the ancestors of the animals now living. Now, fossil forms are unlike existing ones, therefore modification, and very considerable modification, must have occurred. Do the fossils themselves show evidence of such modification? Can we point to a series of forms showing progressive modification towards the present condition? I admit at once that fossils do not give us all the evidence we could wish for; in some instances a fairly complete series can be pointed out, but in most cases we are unable to do this. This is undoubtedly at first sight a serious check, and is one often referred to. By Darwin himself it was stated to be the difficulty that would probably be most widely felt. This difficulty must be considered fully and from two main standpoints : First, of what nature is the record yielded by fossils, and how far is it reasonably complete ? Secondly, are we quite certain that we know in what direction to look for intermediate forms, and are we clear that we should recognise such if we found them ?

The geological record is imperfect for the following reasons :—

1. Only certain parts of certain animals can be preserved as fossils.

Among PROTOZOA the *Foraminifera* and *Radiolaria* are well preserved, *Infusoria* not at all. In PORIFERA the skeletons are well preserved, and of these there is a long record. Of CŒLENTERATA such forms as *Hydra*, jelly-fish, and sea-anemones cannot be preserved, save very exceptionally ; but corals are peculiarly suitable, and few classes are so well represented in a fossil state. ECHINODERMATA are well represented, excepting the *Holothurians*. Among VERMES there is no trace of flat or round worms, but of *Annelids* the jaws, tubes, and tracks are found. ARTHROPODA are well preserved, especially the *Crustacea*, and as a general rule aquatic forms are much more completely preserved than terrestrial ones. In MOLLUSCA the shell is well adapted for preservation, but the air-breathing terrestrial forms are rare. Among the VERTEBRATA we find the bones, teeth, and footprints preserved. Mammals, being terrestrial, are at a disadvantage. Birds, whose bodies are light enough to float, are probably devoured as food, and have less chance of preservation; they are therefore rare as fossils.

These parts can, as a rule, only be preserved in a fragmentary condition.

2. Only certain deposits can so preserve them, notably mud.

3. Animals must die at such places and times that they can be preserved; bones, &c., must be carried down regularly to a particular locality.

4. The deposit must extend over a very long time and continuously, if the series is to be complete, or even approximately so.

5. The area must be in subsidence or else it would be filled up.

6. Fossils once imbedded must be raised above the sea.

7. They must escape denudation and be exposed at some workable spot.

8. The intervals between the deposits often represent times of denudation. The denuded parts will be the uppermost—*i.e.*, the most recent, and will break the series effectively by the removal and destruction of fossil records.

All these conditions can very seldom be fulfilled. The difficulty is well illustrated by our state of knowledge with regard to domestic animals. Where are the bones of the intermediate forms between the rock-pigeon and the pouter, or fantail, for example, to be found? Yet we know that these existed but a short time ago. Speaking of this point, Darwin says : "We shall perhaps best perceive the improbability of our being enabled to connect species by numerous fine intermediate fossil links, by asking ourselves whether, for instance, geologists at some future period will be able to prove that our different breeds of cattle, sheep, horses, and dogs are descended from a single stock or from several aboriginal stocks. . . . . This could be effected by the future geologist only by his discovering in a fossil state numerous intermediate gradations ; and such success is improbable in the highest degree."

It is irrational to demand a perfect gradational series in any number. The actually preserved record is well described as a "chapter of accidents." We have no right to expect, in any particular case we choose to select, that the chain shall be complete, and the links all forthcoming on demand; but we have a right to expect some few well-marked examples of transitional series of forms, and also that none of the facts actually ascertained shall be inconsistent with our theory.

The Geological Evidences of Evolution.

The second preliminary point concerns the nature of fossils. Directly intermediate forms between two existing genera must not be expected, and can indeed very rarely have existed. The theory of Evolution requires that distinct genera shall be linked together, not by a direct connection, but by the descent of both from a common ancestor.

Wallace says: "The fantail and pouter pigeons are two very distinct and unlike breeds, which we yet know to have been both derived from the common wild rock-pigeon. Now, if we had every variety of living pigeon before us, or even all those which have lived during the present century, we should find no intermediate types between these two—none combining in any degree the characters of the pouter and fantail. Neither should we ever find such an intermediate form, even had there been preserved a specimen of every breed of pigeon since the ancestral rock-pigeon was first tamed by man."

This point may be illustrated by an example taken from another department of knowledge—the science of language. Take a word such as *regnum*, and see how the word has become modified in the different languages of European nations, when exposed to different conditions of environment. The following table shows various modifications derived directly or indirectly from the original word *regnum*.

REGNUM.

REINO .	*Portuguese.*
REINADO	*Spanish.*
REGNO .	*Italian.*
RÈGNE .	*French.*
REIGN .	*English.*
REGIERUNG .	*German.*

The word *regnum* corresponds to the blue rock-pigeon in that it is the parent of all the other forms of words, and represents a fossil form or extinct word, belonging to a dead language.

Again, to return to biology, *Phenacodus* (Fig. 6), one of the most important of recent fossil discoveries, was found in the Eocene of North America, and in several forms, varying in size from that of a small terrier to a leopard. This is a good example of a generalised type, having five clawed digits, a small brain, and complete radius and ulna. In many ways it suggests the ancestral form from which the

Fig. 6.

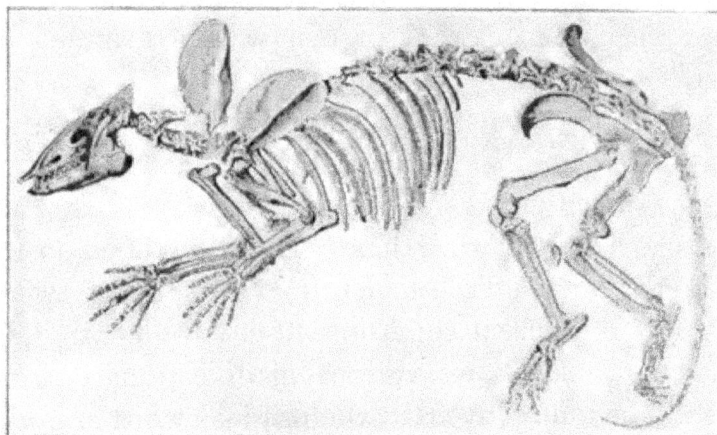

Phenacodus.

Artiodactyla (deer and sheep) and the *Perissodactyla* (rhinoceros and horse) may have sprung; perhaps, also, it is nearly the parent form of the Carnivora. Among *Birds and Reptiles* a well-known series is known, due to Professor Huxley, connecting the one group with the other. Birds form a very compact and sharply limited group, characterised by their wings, feathers, and other peculiarities.

In the first edition of the "Origin of Species," Darwin said: "We may thus account even for the distinctness of whole classes from each other—for instance, of birds from all other vertebrated animals, by the belief that many animal forms of life have been utterly lost, through which the early progenitors of birds were formerly connected with the early progenitors of the other vertebrate classes."

This was a prophecy out of which much capital was made at the time. It appeared an easy way out of the difficulty to suppose extinction and disappearance of all those forms whose existence, at one time or another, it was necessary to assume. At the time of the utterance of this prophecy, in 1859, there was no positive evidence at all. But in 1862, the *Archæopteryx* (Frontispiece) was shown to be a true bird as regards its feathers and wings, combined with several Reptilian characters, such as the long tail, the nature of the hip-bones, legs, and vertebræ. A second specimen was found in 1879, having a skull with numerous teeth, clawed fingers, perfect feathers, and bi-concave vertebræ. In 1868 Professor Huxley showed in fossil reptiles (*Dinosaurus*) the nature of the modification in virtue of which the quadrupedal reptile passed into the type of a bipedal bird. Again, in 1875, the discovery of toothed birds in chalk by Marsh completed the series of transitional forms between birds and reptiles. From that time Darwin's prophecy could be replaced by demonstrated facts. There are actual fossils which bridge over the gap between reptiles and birds, in this sense that they enable us to picture to ourselves forms from which both birds and reptiles as we know them could have sprung. The *Pterodactyl* shows how flight is possible to a reptile, and is possibly related to birds, although this point is doubtful. (Fig. 7.)

The most famous instance of geological evidence

FIG. 7.

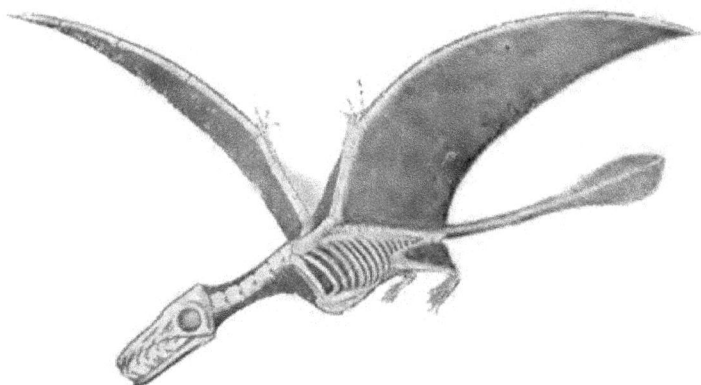

Flying Reptile (Diagrammatic Figure).

is found in the Horse, and, although familiar, is so important as to bear repetition. The typical number of toes and fingers is five, as in ourselves. In quadrupeds generally the number is reduced, but the horse, zebra, and ass stand alone in having only one digit on each foot, corresponding to the middle finger and toe. If we compare the foot of the horse with that of man, we find the "hock" of the horse corresponds with man's heel; the "cannon-bone" is the metacarpal; the "pasterns" form the first two

THE ARGUMENT FROM PALÆONTOLOGY   67

phalanges, and the "coffin-bone" the terminal
phalanx of the toe. The "hoof" corresponds to the
nail. In the fore-limb the "knee" of the horse is
equivalent to the wrist. The "splint-bones" represent the metacarpal bones of the first and third
digits.

Now the ancestors of the horse are *Protohippus*
or *Hipparion*, which is found in the Pliocene;
Miohippus and *Mesohippus*, found in the Miocene;
Orohippus, in the Eocene; and *Eohippus*, at the base
of the Eocene. In *Protohippus* each foot has three
well-formed digits; *Miohippus*, in addition to this,
has a rudimentary metacarpal bone of a fourth digit
in the fore-foot; in *Mesohippus* this rudimentary
metacarpal bone is more fully developed; in *Orohippus* there are four well-developed digits in the
fore-foot, three in the hind-foot; while in *Eohippus*
five digits are present. Thus this series of fossil
forms furnishes a complete gradation from the older
tertiary forms with four or five toes up to the horse
with one toe. These forms differ not only as regards
the number of toes, but also in other respects, chiefly
in the gradual diminution and loss of independence
of the ulna and fibula, and in the gradual elongation
of the teeth and increasing complexity of their
grinding surfaces.

An excellent series of gradational forms is shown
in the case of *Paludina*, of which six or eight disconnected forms were known first, and described as
distinct species; later on, connecting forms were
discovered, and it was realised that we had a case of
progressive modification from the older geologic beds

to the newer ones, and all forms are now included
as varieties of one species.   Over 200 varieties
have been discovered in enormous numbers; one
characteristic form being found in each horizon.
The simpler unornamented shells are from the
lowest layers; the most recent forms being identical
with a form now living only in North America and
the fresh-water lakes of China, which was formerly described as a distinct genus.   This evidence
was found since the publication of the "Origin
of Species" in 1859, and renders the record less
incomplete.

Now that men realise the value of Palæontology,
more attention has been directed to the subject; for
in all cases positive palæontological evidence may be
implicitly trusted, although negative evidence is
worthless.

THE EXTINCTION OF SPECIES.

When a species or group has once disappeared
there is no reason to suppose that the same identical
form ever reappears—*i.e.*, its existence, so long as it
lasts, is continuous.

The influence of the size of animals, and its bearing
on Extinction of Species, is of the greatest possible
interest and importance.   Many zoologists hold the
view, in support of which evidence is steadily increasing, that the primitive or ancestral members of
each group were of small size.   Thus, in the case of
birds, on the whole small birds show more primitive
conditions of structure than the larger members of

the same group, and the first birds were probably smaller than Archæopteryx. Reptiles and
mammals also show in their earlier and smaller
types more primitive features than their larger
descendants.

Again, in the pedigree of the horse, one of the
most striking points is the progressive reduction in
size met with as we pass backwards in time.   The
Pliocene *Hipparion* was smaller than the existing
horse ; the Miocene *Mesohippus* was about the size
of a sheep ; while the Eocene *Eohippus* was no larger
than a fox.   Not only is there good reason for
holding that, as a rule, larger animals are descended
from ancestors of smaller size, but there is also much
evidence to show that increase in size beyond certain
limits is disadvantageous, and may lead to destruction rather than to survival.   It has happened
several times in the history of the world, and in
more than one group of animals, that gigantic stature
has been attained immediately before extinction of
the group, a final and tremendous effort to secure
survival, but a despairing and unsuccessful one.
The Ichthyosauri, Plesiosauri, and other extinct
reptilian groups, the Moas, and the huge extinct
Edentates, are well-known examples ; to which
before long will be added the elephants and the
whales.

The same classification applies to both recent and
fossil animals ; the large divisions are the same, but
many minor groups have become extinct.   All
existing groups are not known to have existed for
all time, and many have certainly not done so.   Still

no one primary division of the animal kingdom is entirely extinct : it is merely the subdivisions that have died out. The earliest origin of all the great groups is driven back to extremely remote times, to the Palæozoic period, and palæontology tells us nothing about the mode of origin of the great divisions of animals. Darwin says : " I look at the geological record as a history of the world imperfectly kept, and written in a changing dialect ; of this history we possess the last volume alone, relating only to two or three countries. Of this volume, only here and there a short chapter has been preserved ; and of each page, only here and there a few lines. Each word of the slowly changing language, more or less different in the successive chapters, may represent the forms of life which are entombed in one con-secutive formation, and which falsely appear to us to have been abruptly introduced. On this view the difficulties above discussed are greatly diminished, or even disappear."

Geographical Distribution.

The explanation of the distribution of animals on land and in the sea is a subject of great importance, which I propose here to touch upon only as it is affected by palæontological evidence.

Much information has been collected on this subject by exploration and by systematic obser-vations obtained by dredging expeditions. A gradually growing conviction has arisen that we must not be content with mere facts, but must demand an explanation of these facts, and that this explanation is in our power to find. It is to Wallace that we are especially indebted for our knowledge of the geographical distribution of animals.

The nature of the problems we have to consider is best shown by examples, of which the following will serve our purpose.

A. CAMELIDÆ, or Camels, are an exceedingly restricted group, the majority of species now living in domestication.

1. *Camelus* is highly characteristic of hot, parched deserts, and is found in Sahara, Arabia, Persia, Turkestan, and Mongolia, as far as Lake Baikal. There are none now living perfectly wild. Of Camelus there are two kinds : the dromedary, found in Asia Minor and Africa, has one hump; the Bactrian camel, possessing two humps, is confined to Asia, and especially Central Asia, north of the Himalayas.

2. *Auchenia* is of smaller size, with slender legs, and has no hump. It is confined to the mountainous and desert regions of the southern part of South America, and is often found on rugged snow-clad slopes at great elevations. Of this group, *Llama* and *Alpaca* are entirely domesticated—the former being used as a beast of burden in Peru and Bolivia ; the latter is cultivated both for its wool and for its flesh. *Vicuna*, the smallest member of the group, is found at elevations of 13,000 feet and upwards, in the Andes of Peru, Ecuador, and Bolivia. *Guanaco*, an animal the size of a fallow-deer, is

distributed in the plains of Patagonia and Tierra del Fuego.

Camels are thus distributed over two areas, comprising the mass of two continents, divided by a great ocean ; one area being north of the equator, the other south of it, and separated by half the circumference of the globe. They are animals of large size, and it is hardly possible for their existence to have been overlooked. Hence we may assume that their geographical distribution is known correctly.

This is a good example of the difficulties in accounting for geographical distribution, and of the way in which they may be met. Evolution tells us that close anatomical resemblances mean near kinship, and forbids us to contemplate the possibility of animals, agreeing in a number of important points, having come into existence independently.

The anatomical characters of camels are well marked; they have two toes—viz., the third and fourth, and walk on the palmar surface of the middle phalanx, not on hoofs. The sole of the foot is formed by broad integumentary cushions, and the nails are small and flattened. The stomach consists of a paunch with smooth lining, provided with two groups of water-cells with narrow mouths. The cervical vertebræ are peculiar, in that the canal for the vertebral artery pierces the arch of the vertebra, instead of the transverse process.

The theory of gradual modification of animals

renders it impossible that identical conditions could have been acquired twice independently. The fact that the two groups of camels agree in a large number of points, in which they differ from all other animals of the same class, must be taken as a proof of near blood-relationship.

Near blood-relationship means common origin— *i.e.*, one of the two groups of camels must be descended from the other, or both groups must be descended from some common ancestors, which were already camels. In other words, either the new-world camels must be descended from the old-world camels, or *vice versâ;* or both must be descended from camels that formerly lived elsewhere, but are now extinct.

Our problem is now becoming more clearly defined, and we have to consider the means of migration of mammals. The only means of migration is by walking; for although most of them can swim, it is only for short distances, and it is doubtful whether any land mammal can swim across an arm of sea fifty miles wide. Captain Webb's swim across the Channel has perhaps never been beaten by a land mammal. The practical proof of the efficacy of the sea as a barrier to migration is seen in the fact of the absence in most oceanic islands of indigenous mammals, except bats. To put it in plain words, if mammals are to get from one place to another, they must walk.

The only explanation possible is through fossils, which thus have a new interest, depending on what parts of the earth we find them in. The evidence of fossils with regard to camels is very imperfect, but still points in a fairly definite direction. Fossil camels are found in South America, in Brazil; in North America, in Texas, California, Kansas, and Virginia. In Asia, in the Himalayas, *Merycotherium*, a large fossil camel, is found widely distributed over Siberia, extending to the extreme coast.

Now, there was almost certainly a former land connection between Asia and North America, across the Behring Straits, which are narrow and shallow. Hence the conclusion is, that there is strong reason for holding that camels originated in North America, and thence spread in two directions, southwards to South America, and westwards through Asia; and that their areas of distribution, though now disconnected, were once continuous.

B. MARSUPIALS.—These constitute a large group of animals, of which there is a great variety of forms, the kangaroo and opossum being perhaps the best known examples. Marsupials are a well-marked group, which in their habits, appearance, and structure, especially as regards the skeleton and teeth, curiously simulate the higher divisions of mammals. Carnivorous, insectivorous, and herbivorous forms are all well established and differentiated. The ant-eater or *Myrmecobius*, and fruit-eater or *Phalanger*, are found in this group, which is characterised by low organisation and great tenacity of life.

Marsupials occur in two chief regions:

(i.) *Australian region.*—The wombats and Myrmecobius occur only in Australia and Tasmania, kangaroos and phalangers extending northward to

New Guinea and adjacent islands; phalangers to Timor, the Moluccas, and Celebes.

(ii.) *American region.*—The opossums are most numerous in the forest region of Brazil, south of the river La Plata; also west of the Andes in Chili. Their distribution extends northwards to Mexico, Texas, and California; and in the States from Florida to the Hudson river, and westwards to the Missouri.

Marsupials form a good example of discontinuous distribution, the explanation of which is yielded by fossils. Opossums are found in the Tertiary deposits of England, France, in other parts of Europe, and in North America. The Cretaceous period shows no trace of them, but in the Jurassic, and in the yet older Triassic, at the base of the Secondary series, large numbers of mammalian remains of small size have been found, which are considered to represent the early phase in marsupial development. Here the starting-point or birth place appears to have been in the Old World, and the group to have migrated southwards.

C. TAPERIDÆ, or tapirs, are found in the equatorial forests of South America, in the Andes of Ecuador, in Panama and Guatemala, and also in the Malay Peninsula, Sumatra, and Borneo. Geological evidence shows that during the Miocene and Pliocene times, tapirs abounded over the whole of Europe and Asia, and their remains are found in the Tertiary deposits of France, India, Burmah, and China. In both North and South America fossil remains of tapirs occur only in caves and deposits of the Post-

Pliocene age, showing that they are comparatively recent immigrants into that continent, perhaps by means of the Behring Straits again. The climate even now is much milder than on the north-east of America, and perhaps was warm enough in late Pliocene times to allow emigration of tapirs, which were driven south to the swampy forests of the Malay region.

CONCLUSION.—On the whole, then, the evidence afforded us by fossils is not so complete as we should wish, and we have seen that from the necessity of the case this must be so. However, evidence is steadily accumulating, and such evidence as we have is not merely favourable, but in some instances remarkably complete. Indeed, since the date of publication of the "Origin of Species," in 1859, our knowledge has increased, and evidence has accumulated so markedly, that it has been said by a highly competent authority, that if the doctrine of Evolution had not existed, palæontologists would have been compelled to invent it as the only possible explanation of the facts determined. Again, we are not aware of the existence of palæontological facts which can be demonstrated to be inconsistent with the theory; while the explanation which they afford of new and previously unstudied problems, such as some of the questions of geographical distribution we have touched upon, is evidence of a strong nature in support of the theory. Professor Huxley, with regard to this subject, says: "The primary and direct evidence in favour of Evolution can be furnished only by palæontology. The geological record, so soon as it approaches completeness, must, when properly questioned, yield either an affirmative or a negative answer : if Evolution has taken place, there will its mark be left ; if it has not taken place, there will lie its refutation."

THE last lecture was devoted to the consideration of the evidence afforded by fossils in regard to the theory of Evolution, and may be summed up as follows :—The evidence afforded by fossils is not so complete as we could wish, but we are able to point out the causes which render it difficult or impossible for continuous series to be preserved ; fossils give no evidence against Evolution, and some remarkable series have already been unearthed, such as those of the Horse and *Paludina*, which would be unintelligible without Evolution; the evidence is steadily increasing in amount and importance ; and the evidence of fossils is a disproof of catastrophism.

We are now concerned with the most recent of biological sciences. Embryology, or the Science of Development, is prominently associated with the names of Von Baer, 1792–1881, and Balfour, 1851–1882. It is utterly vain in one lecture to give any idea of the extraordinary multitude of facts accumulated within the last quarter of a century, or of the numerous and fascinating theories to which these facts have given origin ; it is here merely as bearing

on the doctrine of Evolution that we have to consider
Embryology.

There are two great questions to be considered :—
First : Does embryology afford evidence for or
against the possibility of the descent of animals from
unlike ancestors ?   Secondly : If it gives evidence in
favour of such descent, does it afford us any clue in
regard to the actual line of descent in a given case,
and will it help us to reconstruct the pedigrees or
past histories of animals ?

The answer to the first question is found in the
extraordinary changes which an animal may undergo
in its own person, during development, within the
space of a few days or weeks, thus showing the possibility of such descent with modification ; for instance,
the changes which occur during the metamorphosis of
the butterfly, and the change of the water-breathing
tadpole into the air-breathing frog.   This suggests
further that such enormous periods of time as are
usually demanded to bring about such changes may
not really be necessary.   A further reply to the
question is found in the fact that groups of animals,
the relations of which were previously unknown,
have had their true zoological positions determined
by the study of the changes undergone during their
development.

Thus many animals, when adult, present little or
no resemblance to other members of the groups to
which they really belong.   The Ascidians are a well-
known example.   So long as their adult condition
alone was known, zoologists were entirely in the
dark as to their real affinities, and by most writers

they were grouped with the *Brachiopoda* and *Polyzoa*, as a subdivision of *Mollusca*. As soon as their development was worked out, it was found that they were really members of the *Vertebrata*, inasmuch as when young they show a curious resemblance

FIG. 8.

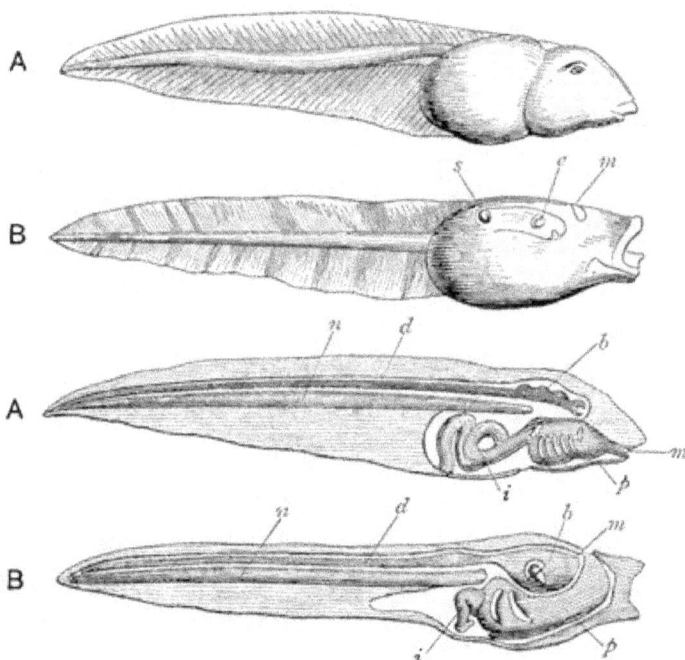

Tadpoles of Frog and Ascidian
(A, Frog ; B, Ascidian).
b, brain ; *d*, spinal cord ; *e*, eye ; *i*, intestine ; *m*, mouth ; *n*, notochord ; *p*, pharynx with gill-slits ; *s*, spiracle

to tadpoles—not merely in form and appearance, but in all essential points of structure ; while the mode of formation of their nervous system, skeleton, alimentary canal, and other parts, is exactly that obtaining among other Vertebrates, and entirely unlike that of all Invertebrate groups. (Fig. 8.)

Again, we may take animals such as a prawn, a barnacle, and one of those curious sac-like parasites of the genus *Sacculina*, which are found not uncommonly adhering to the under surface of the rudimentary tail of crabs.   These three animals are, when adult, very unlike one another.   The prawn is a free-swimming form.   The barnacle is fixed firmly to rock,

Fig. 9.

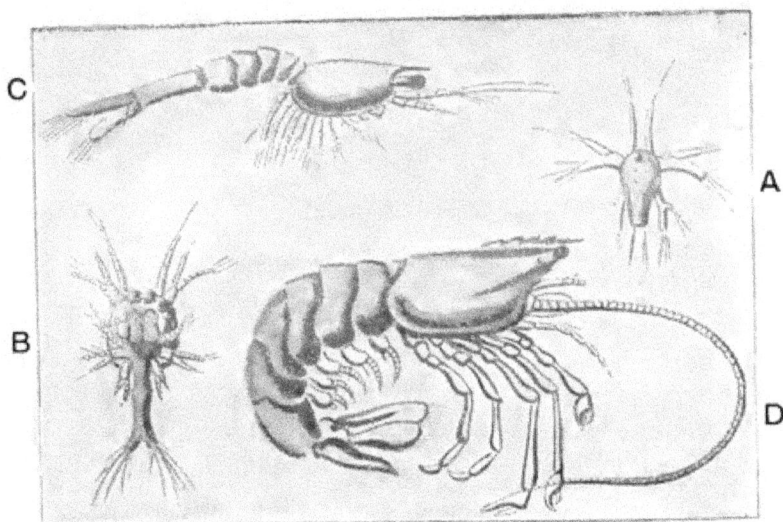

Stages in Development of the Prawn (Peneus).

A, *Nauplius* stage ; B, *Zoœa* stage, in which the larva resembles an dult Copepod ; C, *Schizopod* stage, where it corresponds in structure to the adult *Schizopoda* ; D, Adult Peneus.

usually between tide-marks ; it forms a hard protective shell, has no eyes, no locomotor organs, and is hermaphrodite.   Sacculina is altogether unlike the other two animals : it has a soft unjointed body, no trace of limbs, no mouth or alimentary canal, and no sense-organs : it is, in fact, merely a soft-walled bag of eggs, attached to the crab's tail by a number of branching root-like

F

processes penetrating the crab's skin and spreading
out in its body, from which they absorb the nutriment on which the parasite lives and grows.

Yet when we turn to the development of these
animals, we find that, utterly unlike as the adult

FIG. 10.

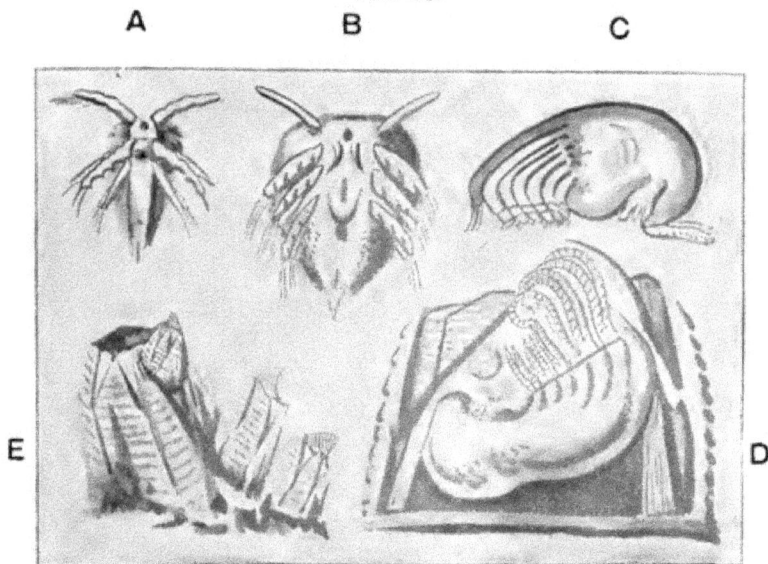

Stages in the Development of the Barnacle (Balanus).

A, *Nauplius* stage; B, Second stage, in which the first pair of
swimming appendages of the Nauplius are converted into antennæ and the
rudiments of the six pairs of cirri appear; C, Pupa stage—in this stage
the animal is free-swimming and has six pairs of legs, antennules, two large
compound eyes, and imperfectly developed masticatory appendages. The
pupa becomes attached by its antennules and develops into D, the adult
Barnacle. E, Group of Barnacle shells.

forms are, the young of all three genera hatch in the
form known as a *Nauplius*. This Nauplius larva
has a short unsegmented body, three pairs of
appendages, used for locomotion, and a single median
eye; and although the Nauplii are not identical one
with another, yet they agree very closely in all

FIG. 11.

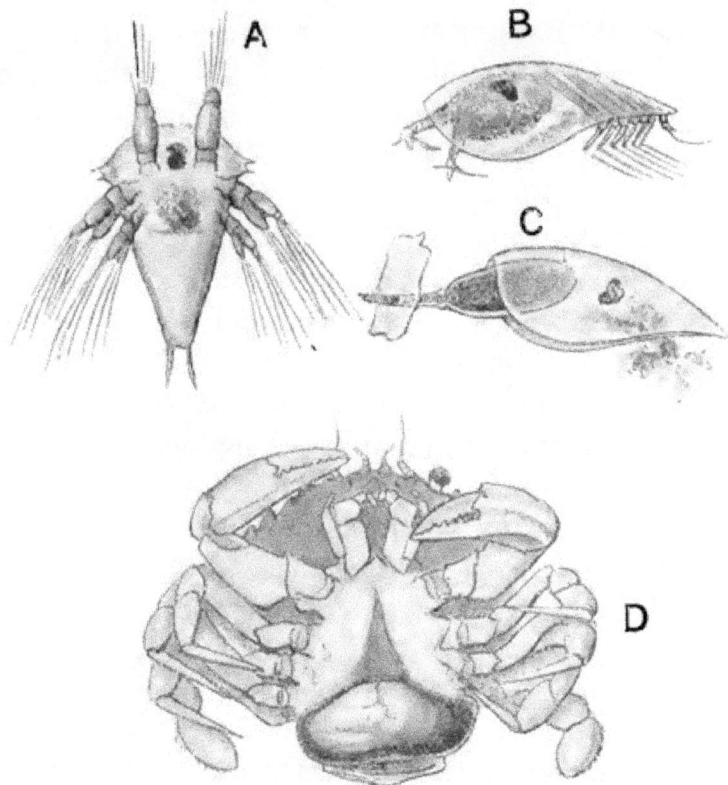

Young and adult specimens of Sacculina, *to illustrate the Degeneration or Retrograde Metamorphosis which the parasite undergoes in the course of its development.*

A.—The Nauplius stage, in which the young *Sacculina* hatches. The three pairs of appendages correspond to the antennules, antennæ, and mandibles of a Crab or Crayfish. The black spot between the antennules is the eye, and the small patch immediately behind the eye and between the two hinder pairs of appendages is the ovary, which is already present at this very early stage. Unlike the Nauplius stage of other groups of Crustacea the *Sacculina* Nauplius has no mouth or alimentary canal. × 90.

B.—The *Cypris* or pupa stage. It is at this stage that the young *Sacculina*, hitherto a free-swimming animal, attaches itself to a Crab and becomes parasitic. The pupa is characterised by the bivalved carapace; the stout antennules by which it fixes itself to the Crab; and the six pairs of locomotor appendages. The black spot is the Nauplius eye, and the mass immediately below it is the ovary. × 90.

C.—The *Sacculina* three days after fixing itself to the Crab. The six pairs of swimming legs have rotted away and fallen off; the bivalved carapace is being detached, and is carrying with it the eye and certain *débris* from the body. The sole parts remaining, out of which the adult *Sacculina* will be formed, are the antennules, now modified into a tube, which is represented projecting through a piece of the skin of the Crab; and the head, which forms a bottle-shaped mass attached to the tube, and containing the ovary. × 90.

D.—Adult *Sacculina* attached to the ventral surface of the tail of a Crab (*Portunus*). The *Sacculina* is the large dark-coloured bag in the lower part of the figure; it is attached to the Crab by a short fleshy stalk, not seen in the figure, which, penetrating the skin spreads out into a tuft of branching roots in the Crab's body. × ½

essential features.   Embryology tells us that this
means that the three animals must really be members
of the same group, and allied to one another ; and
thereby gives us a clue to the real affinities of barnacles and Sacculina that we could hardly get in any
other way.   (Figs. 9, 10, 11.)

In parasitic animals generally the shape and structure are liable to be so profoundly modified, in consequence of the special conditions of parasitic existence, that, but for the aid afforded by development,
we should often be absolutely unable to determine
to which division of the animal kingdom they really
belong.

This leads us to our second question : If Embryology gives the clue to the relationship of animals,
may it not do more and reveal their ancestry ?

THE RECAPITULATION THEORY.

A further explanation is afforded us by what is
known as the *Recapitulation Theory*, which states that
not merely have existing animals descended from
ancestors which are often unlike them, but that
each animal bears the mark of its own ancestry and
reveals its parentage in its own development.   Evolution tells us that each animal has had a pedigree
in the past ; Embryology reveals to us this ancestry,
because every animal has an inherited tendency
during its own development to repeat its own ancestral history ; or, to put it in other words, to climb
up its own genealogical tree.

A good example of recapitulation is afforded by

flat fish such as the sole, flounder, turbot, and plaice, which are distinguished not merely by the remarkable flattening of the body from side to side, but by the further facts—(1) that the two sides, right and left, of the fish are never coloured alike, one being nearly white, and the other dark-coloured ; and (2) that the two eyes, instead of being situated, as in other animals, one on each side of the head, are both on the same side—*i.e.*, the darkly-coloured one.   On watching these flat fish in an aquarium, we note that they habitually lie on the bottom on the paler coloured side, and we are at once led to associate the remarkable condition of the eyes with this habit ; for it is clear that when so resting, if the eyes were placed in the usual positions, one at each side of the head, the eye on the paler surface—*i.e.*, the surface on which the fish lies—would not only be perfectly useless, but would be liable to injury from contact with the sea-bottom.

On turning to the development of the flat fish we find this supposition confirmed.   A sole on hatching, and for some time afterwards, has its eyes one on each side of the head, just like any ordinary fish ; furthermore, it swims, like other fish, with the body vertical, and has its two sides coloured alike.   It is only after it has attained some size that it gradually adopts the habits of the adult, and takes to resting on its left side on the sea-bottom and swimming with the body horizontal instead of vertical.   At the same time, the right side of the body gradually becomes coloured differently to the left, and in such a way as to resemble the sea-bottom closely, and so enable the

fish to escape the notice of its enemies ; and the left
eye, no longer of use in its original position, is

FIG. 12.

A

B

C

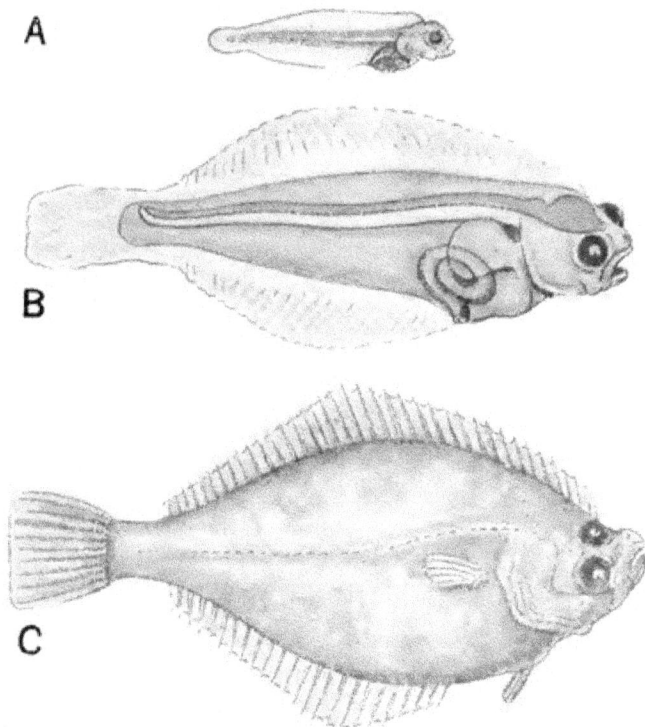

Young and adult specimens of the Flounder (Pleuronectes flesus), to illustrate the shifting of the eye from one side of the head to the other during the growth of the fish.

A.—A young Flounder six days after hatching. The head is symmetrical, and the eyes one on each side. ×4

B.—A young Flounder, probably about a month old. The fish is gradually acquiring the characteristic shape of the adult, and the head is becoming twisted, the eye of the right side being displaced slightly downwards, and the eye of the left side coming into view over the top of the skull. ×4

C.—An adult Flounder showing the characteristic shape of the flat-fish, and the complete migration of the left eye over to the right side of the head. ×½

gradually displaced upwards on to the top of the
head, and then shifts over to the right side ; the

change in position of the eye being accompanied by very considerable twisting and distortion of the skull. (Fig. 12.)

Inasmuch as flat fish in all other respects agree with more ordinarily-constituted fish, and as their special peculiarities—*i.e.*, the lateral compression of the body, the difference of colouring on the two sides, and the singular position of the eyes—may all be readily and naturally explained by their habits, which again would be clearly advantageous to the fish in aiding them to escape from enemies, it becomes in the highest degree probable that flat fish are descended from normally-formed fish, which first acquired the habit of lying on one side for the sake of protection, and then underwent structural changes in consequence of this habit.

If this be correct, then the developmental history of the flat fish becomes intelligible by assuming that each individual has an inherited tendency to repeat in its own development the history of the species ; every flat fish during its own growth passing through the same series of changes by which we have supposed the whole race of flat fish to have acquired their special peculiarities.

The case with regard to the sole is really a very strong one : for the only alternative view is that flat fish are not descended from normally-shaped fish, but have sprung into existence independently ; and not only is this view absolutely contradicted by what we know of other animals, but it would render the development of the flat fish an incomprehensible mystery. The one view gives a

complete and intelligible explanation of all the facts of
the case; the other not only has no direct evidence
in its favour, but is totally opposed to all experience,
and leaves the developmental features unexplained
and inexplicable.

I have selected the sole for special description
because the facts of the case are well known, and the
argument is a simple and easily-followed one.
Other animals, however, would serve the purpose
equally well, and would afford illustrations quite as
striking of the aid given us by the Recapitulation
theory in unravelling embryological problems.

Thus, a crab and a lobster are animals closely
agreeing with one another in essential structure, and
clearly belonging to the same zoological group.
The characteristic difference in form between the
two is due to the fact that, while in the lobster the
hinder part of the body, or "tail," is well developed,
forming about half the length of the animal, and
being used as a swimming organ; this "tail" is in
the crab very greatly reduced in size, is of no use
for swimming, and instead of projecting horizontally
backwards, is carried bent forwards under the
anterior part of the body, to which it is so closely
fitted as to escape notice at first sight. Here again,
as in the case of the flat fish, the structural differences
may clearly be traced to difference in habit. Lobsters
not merely walk about on the sea-bottom on their
legs, but are able to swim freely in a backward
direction, by powerful jerks of the tail. Crabs, on
the other hand, walk but do not swim; and in them
the tail, being no longer of use, has greatly dimin-

ished in size, and become rudimentary. Crabs,
however, in their early stages of development, are
free-swimming animals, and have tails quite as well
developed and fully as large, relatively to the whole
animal, as lobsters; and it is only after they have
reached a certain size that they abandon their free-
swimming habits, sink to the bottom, and henceforth
move by walking only.   The case is exactly parallel
to that of the flat fish; and the Recapitulation
theory explains the developmental history of a crab
by saying that it is a repetition of the ancestral
history of crabs in general : that crabs are descended
from animals essentially similar to lobsters—*i.e.*,
from *Macrurous* ancestors, and that each crab
passes through a lobster stage in its development, because of the inherited tendency that all
animals have to climb up their own genealogical
trees. (Fig. 13.)

The evidence of the descent of crabs from Macrurous ancestors involves, moreover, the supposition
that they came into existence later than the *Macrura*.
This supposition is supported by the evidence of
palæontology, for the *Macrura* are found as fossils
in the Devonian and Carboniferous periods, and
abundantly so in the Jurassic and Cretaceous, but
are comparatively scanty in the Tertiary period ; the
Brachyura, or crabs, on the other hand, are very
abundant in the Eocene and numerous in the
Cretaceous, but doubtfully represented in the earlier
periods.

Good examples of recapitulation are found in
Molluscs. The typical Gasteropod has a large

spirally-coiled shell; the Limpet, however, has a
conical shell, which in the adult animal shows no

FIG. 13.

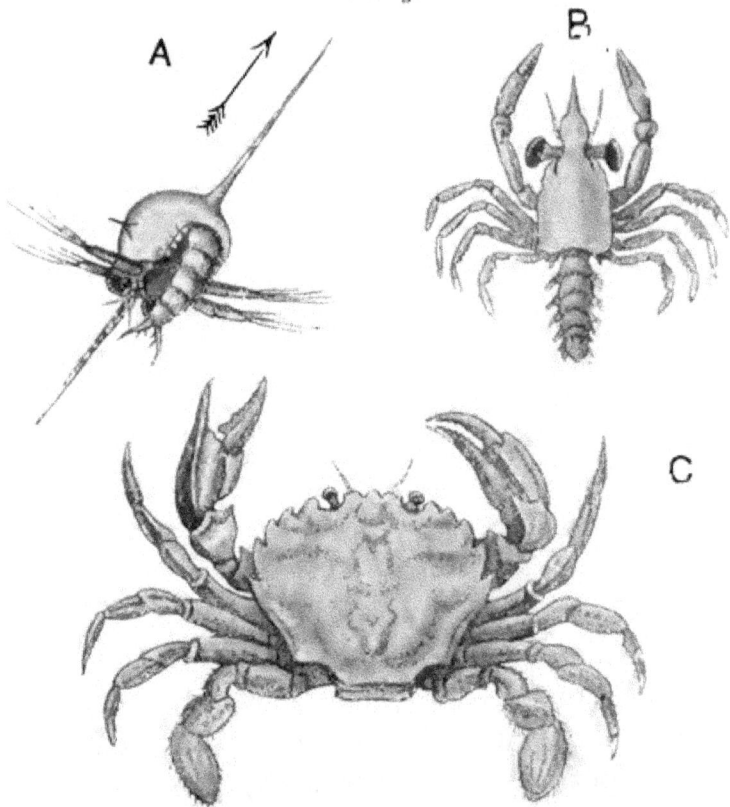

*Young and adult specimens of one of the swimming Crabs (Portunus), to
illustrate the transition from the long-tailed to the short-tailed condition.*

A.—The Zoœa stage, characterised by the great length of the spines on
the cephalothorax; by the large size of the powerful rowing maxillipedes;
and by the long, jointed tail. The larva is swimming in the direction of
the long dorsal spine, the spines serving to guide its course.   × 15

B.—The Megalopa stage. This is the typical Macrurous condition,
comparable to that of an adult Lobster or Prawn.   × 5

C.—The adult Crab. As compared with the Megalopa stage the
cephalothorax has increased greatly in width, while the tail has become
relatively smaller, and is carried turned forwards beneath the thorax. × ½

sign of twisting, although the structure of the animal
shows its affinity to forms with spiral shells. How-

ever, in its early stages of development the Limpet has a spiral shell, which is lost on the formation of the conical shell of the adult.

Recapitulation is not confined to the higher groups of animals, and good examples are found among the Protozoa. One of the best instances is that of *Orbitolites*, one of the most complex of the Foraminifera, which, during its own growth and development, passes through the series of changes by which the discoidal type of shell is derived from the simpler spiral shell. This forms an instructive example, for, owing to the mode of growth by addition of new shelly matter, the older parts are retained often unaltered, and in favourable examples all stages can be determined by simple inspection of the adult shell. (Fig. 14.)

The mode of growth of shells is important, since it gives an opportunity for *comparing the palæon-tological and embryological records*. In such a shell as that of the *Nautilus* the central chamber is the oldest and first formed one, to which the other chambers are added in succession. If then the development of the shell is a recapitulation of ancestral history, the central chamber should repre-sent the palæontologically oldest form, and the remaining chambers in succession forms of more and more recent origin.

In the shells of *Ammonites* it has been shown that such a correspondence between historic and embryonic development really exists. In the middle Jurassic deposits the older Ammonites are flattened and disc-like, with numerous ribs ; in later forms the

FIG. 14.

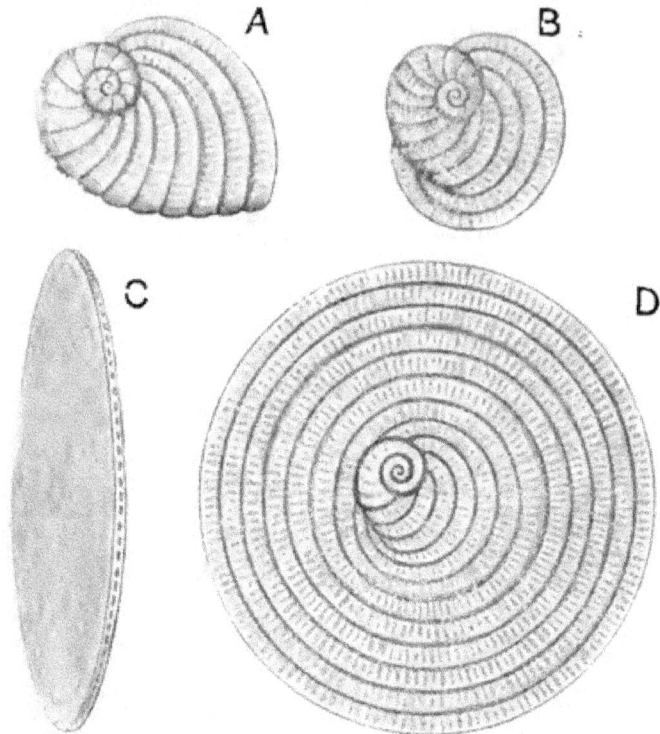

Shells of Peneroplis and of Orbitolites, members of the group of Porcella-
nous Foraminifera, illustrating the mode of transition from
the spiral to the discoidal shell.

A.—An adult *Peneroplis* shell. The shell is spiral and chambered, the
later-formed chambers being very wide, and having a tendency to overlap
the preceding ones.   × 20

B.—A young *Orbitolites* shell, in the *Peneroplis* stage of development.
The shell is spiral and chambered, the last-formed chambers having a
more marked tendency to overlap the preceding ones than in *Peneroplis*.
The last or marginal chamber in the specimen figured extends almost the
whole way round.   × 30.

C.—An adult *Orbitolites* shell, seen edgeways, so as to show the thickness
of the disc, and the marginal pores through which the pseudopodia are
protruded during life.   × 7

D.—An adult *Orbitolites* shell, seen full face. In the centre is the spiral
nucleus, which is the oldest part of the shell, and was originally the only
part present ; then comes a part in which the successive chambers become
wider and wider, and overlap the older part of the shell more and more
completely, and finally the marginal and latest formed part, in which each
chamber is circular and completely surrounds its predecessors.

The fainter radial lines indicate the secondary partitions by which the
chambers are subdivided.   × 7

shell bears a row of tubercles near the outer side of the spiral, and later still a second inner row of tubercles as well, while the ribs gradually become less conspicuous, and ultimately disappear. In more recent forms the outer row of tubercles disappears, then the inner row, the shell becoming smooth, swollen, and almost spherical. On taking one of these smooth spherical shells, such as *Aspidoceras cyclotum*, and breaking away the outer turns of the spiral so as to expose the more central and older turns, first an inner and then an outer row of tubercles appear, which nearer the centre disappear, and in the oldest part of the shell are replaced by the ribs, characteristic of the earlier, and presumably ancestral forms.

Another illustration of the parallelism between the palæontological and the developmental series is afforded by the antlers of deer, which are shed annually, and grow again of increased size and complexity in each succeeding year. In the case of the red-deer (*Cervus elaphus*), the antlers are shed in the spring, usually between the months of February and April; during the summer the new antlers sprout out, and growing rapidly, attain their full size at the pairing season in August or September; they persist through the winter, and are shed in the following spring. The antlers of the first year are small and unbranched; those of the second year are larger and branched; in the antlers of the third year three tynes or points are present; in the fourth year four points, and so on until the full size of the antler and the full number of points are attained.

The geological history of antlers is of great interest. In the Lower Miocene and earlier deposits no antlers have been found. In the genus *Procervulus*, from the Middle Miocene, a pair of small, erect, branched, but non-deciduous antlers were present, intermediate in many respects between the antlers of deer and the horns of antelopes. From slightly later deposits a stag (*Cervus dicrocerus*) has been found with forked deciduous antlers, which, however, do not appear to have had more than two points. In upper Miocene times antlers were more abundant, larger, and more complex; while from Pliocene deposits very numerous fossils have been obtained, showing a gradual increase in the size of the antlers and the number of their branches, down to the present time.

Antlers are therefore, geologically considered, very recent acquisitions : at their first appearance they were small, and either simple or branched once only; while in succeeding ages they gradually increased in size and in complexity. The palæontological series thus agrees with the developmental series of stages through which the antlers of a stag pass at the present day, before attaining their full dimensions.

The Recapitulation Theory, if valid, must apply not merely in a general way to the development of the animal body, but must also hold good with regard to the formation of each organ and system, and with regard to the later as well as the earlier phases of development. Take for example the mode of renewal of nails and of the epidermis generally in the human skin. Each cell begins its career in an indifferent condition, and gradually acquires the adult peculiarities as it approaches the surface, through removal of the cells lying above it.

RUDIMENTARY ORGANS OR VESTIGES.

The Recapitulation Theory also affords most valuable assistance in determining the meaning of special points in the structure of animals which otherwise would be hard to understand. More especially is this the case with regard to what are spoken of as "rudimentary organs." These are structures, such as the eye of the mole, or the rudimentary teeth present in whalebone whales at an early developmental stage, but got rid of before birth : structures which are constantly present in all members of the species, but which are of no use to their possessors. They cannot be nascent structures—*i.e.*, ones which are in process of formation, but which have not yet become functionally active, for the law of natural selection requires that no structure can be developed and retained by a species unless it either is of direct use to its possessor, or else has been of use to some of its ancestors. Their presence would be a complete enigma, but for the Recapitulation Theory, which explains them as structures which were formerly of use to the ancestors of the existing animal, and which appear in the latter because of the inherited tendency of all animals to repeat their ancestral history in their own development. Eyes are developed in the mole, although quite useless to it, because moles are descended from mammals in which the eyes were functionally active. Through the burrowing habits of the mole, the eyes have become degenerate and rudimentary ; but owing to the law of Recapitulation, they are still developed. So with whalebone whales, which are toothless when adult, the presence of teeth at an early stage of development can only be explained by the descent of whalebone whales from toothed ancestors.

Rudimentary organs are of exceedingly common occurrence ; indeed there are probably few if any of the higher animals in which some may not be found. Thus the splint-bones of a horse's leg are rudiments of the metacarpals or metatarsals of the second and fourth digits of the manus or pes, which were fully developed in the extinct *Hipparion*, and in other more remote ancestors of the horse. Almost all parasitic animals undergo, as we have seen, retro-gressive or degenerative change in certain parts or the whole of their structure, and it commonly happens that vestiges of these lost organs linger on as rudiments, whose presence would be inexplicable but for the history of their formation.

Man himself is no exception to the rule. The muscles of the ear, whereby it can be pulled upwards or twitched forwards or backwards, are in a degener-ate condition, and comparatively few men have any real power over them ; while other smaller muscles which run from one part of the ear to another, are in such a completely rudimentary state, that but for anatomy their presence would never be suspected.

There are other parts of man's bodily structure—

his teeth, for example—that show degeneration quite
as clearly as do these ear-muscles ; while most
excellent examples are furnished by his laws, habits,
clothing, and speech.   In such a word as "reign,"
the letter "g" is mute and rudimentary ; no attempt
is made to sound it when pronouncing the word, and
its presence can only be explained in accordance
with the laws of rudimentary organs generally.
Turn to the past history of the word, refer to its
ancestors, and you find in the Latin "regnum" a
word in which the "g" has full value, and from
which we know that our own "reign" has been
derived. The "b" in "doubt," the "n" in "solemn"
are other examples; while in the "lf" of "half-
penny" we have a case in which degeneration is in
the act of taking place at the present time.

Disturbing Causes hindering Recapitulation.

We must now turn to another side of the question.
Although it is undoubtedly true that development is
to be regarded as a recapitulation of ancestral phases
and that the embryonic history of an animal presents
to us a record of the race history; yet it is also an
undoubted fact, that the record so obtained is neither
complete nor straightforward.

It is indeed a history, but a history of which
entire chapters are lost, while in those that remain
many pages are misplaced and others are so blurred
as to be illegible ; words, sentences, or entire
paragraphs are omitted, and, worse still, alterations
or spurious additions have been freely introduced

by later hands, and at times so cunningly as to defy detection.

The chief disturbing cause arises from the necessity of supplying the embryo with nutriment. This acts in two ways. If the amount of nutritive material in the egg is small, then the young animal must hatch early, and in a condition in which it is able to obtain food for itself. In such cases there is of necessity a long period of larval life during which natural selection may act so as to introduce modifications of the ancestral history, or spurious additions to the text. If, on the other hand, the egg contains a considerable quantity of nutrient matter, then the period of hatching can be postponed until this has been used up. The consequence is that the embyro hatches at a much later stage in its development, and if the amount of food material, or food yolk as it is called, is enough, may even leave the egg in the parent form.

This varying condition, as regards amount of food-yolk, affects recapitulation—*i.e.*, the tendency of the embryo to pass through the ancestral stages—in two principal directions. If there is very much food-yolk, there is a tendency for the embryo to shorten its development by the *omission of certain of the ancestral stages*, and especially by the suppression of characters which, though functional in the ancestors, are of no use in the adult state of the animal itself. Thus tadpoles, after hatching, breathe for a time by gills : this gill-breathing condition being an ancestral one for all Vertebrates. In the West Indies there is a little frog (*Hylodes*) which lays its eggs, not in water, but on the leaves of plants. These eggs are larger than those of the common frog—*i.e.*, contain more food-yolk—and the young embryo is thereby enabled, just like the lobster or crayfish, to develop to a later stage before hatching : it passes through the tadpole stage within the egg, and hatches, like the crayfish, in the form of its parent. Although it passes through a gill-cleft stage no gills are developed; being of no use to the embryo, it would be a sheer waste of time to form them, and so they have dropped out of the ontogeny or individual development. (Figs. 15, 16.)

Exactly the same thing has happened in reptiles, birds, and mammals, in which gill-clefts are found, but gills are not developed. A similar tendency to the omission or blotting out of useless characters is seen in the development of all forms which have sufficient food-yolk to carry them over the stages at which these characters would be of functional value before the time of hatching.

In the case of embryos developed from small eggs there is a *tendency to distortion of the ancestral history* of a very different nature. Such embryos, owing to the small supply of food-yolk in the egg, have to hatch very quickly—*i.e.*, not merely of small size, but in a condition representing a very remote ancestral stage. The intervening stages between the early condition and the adult one have to be repeated while the larva is enjoying a free existence : this process will necessarily be slow, for the larva has not merely to develop, but has to obtain for itself food, at the expense of which the further development

FIG. 15.

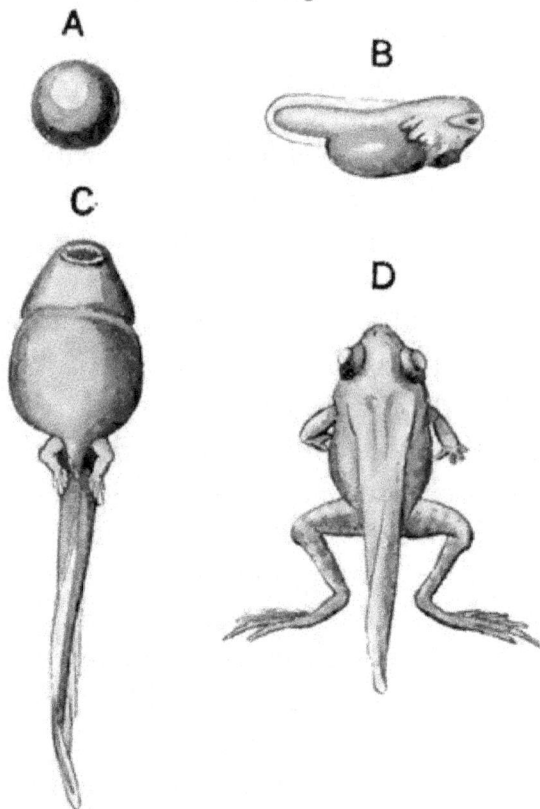

Stages in the development of the common Frog (Rana temporaria).

A.—The egg.   × 3

B.—The Tadpole at the time of hatching. The mouth is not yet formed, the Tadpole being still dependent on the food-yolk present in its body. Two pairs of external gills are present as branched finger-like processes at the sides of the neck. Below and in front of these is the horse-shoe shaped sucker by which the Tadpole fixes itself, and at the sides of the front of the head the rudiments of the nose and eye are seen.   × 3

C.—A Tadpole shortly after the time of appearance of the limbs: the hind limbs are seen at the junction of the body and tail : the fore limbs are present, but are concealed by the opercular folds covering the gills. At this stage the Tadpole breathes by both gills and lungs.   × 1

D.—A young Frog with the tail only partially absorbed.   × 1

FIG. 16.

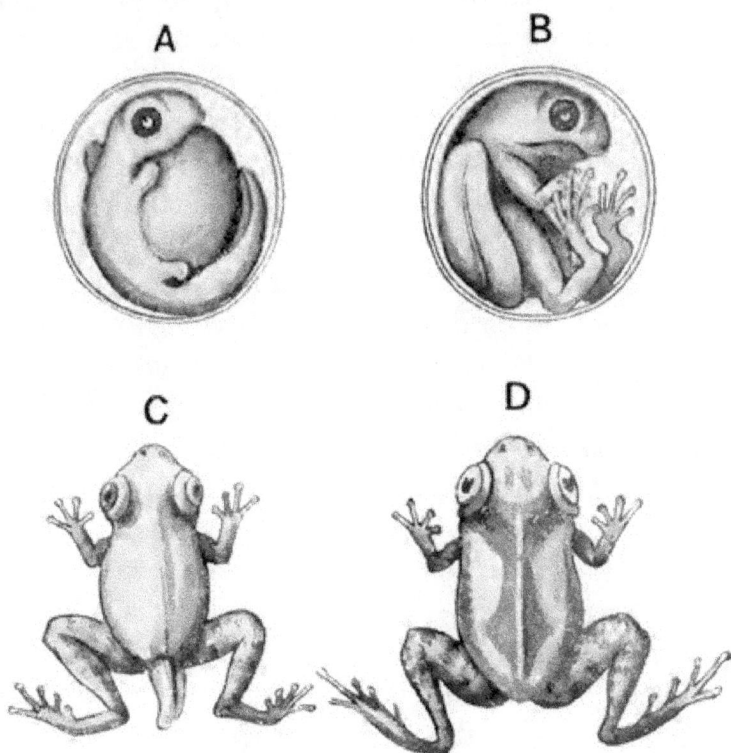

Stages in the development of the West Indian Frog (Hylodes): *illustrating the effect of increased amount of food-yolk in causing the omission of ancestral stages. The free living Tadpole stage of the common Frog is entirely suppressed, and no gills are ever formed. The entire development from the laying of the eggs to the hatching of the Frogs occupies from a fortnight to three weeks.*

A.—The larva at the end of the first week. The head, eyes, stumps of the limbs, and the long tail are well shown. The food-yolk is contained within the large yolk-sac in the middle of the figure.   × 3

B.—The young *Hylodes* shortly before hatching. There is still a very large tail present, which is believed to be used as a respiratory organ.   × 3

C.—Young *Hylodes* at the moment of emerging from the egg : a short stump of a tail is still present.   × 3

D.—Young *Hylodes* at the end of the first day. The tail is completely absorbed, and the Frog has already the form of an adult.   × 3 may be effected.   Hence such larvæ may take weeks, months, or even years in recapitulating these later stages, which their larger-egged allies get through in a few days.   Moreover, during the whole of this time they are exposed to competition, both amongst themselves and with other animals; they have to obtain food for themselves, and they are liable to be themselves devoured as food by other animals.

Owing to this competition, and to the length of time during which it lasts, these *larvæ are liable to acquire, through natural selection, characters which are connected with their existence as larvæ, but which form no part of the ancestral history*; characters which will aid them in obtaining food, or in escaping from their enemies, but which were not found in any of the ancestors of the species.   Of such secondary larval characters the long spines with which the *Pluteus* larva of sea-urchins are provided are good examples; so also are the enormous spines on the young larvæ of crabs and other crustacea.   Other excellent illustrations are afforded by the developmental history of many fresh-water forms, in which, from the danger of their being swept down by the currents of the rivers or streams in which they dwell, or to obtain protection from the cold of winter, special characters are often acquired.   The *glochidium* larva of the fresh-water mussel is a good instance of the former, and the specially-protected statoblasts or winter buds of *Polyzoa* and sponges, of the latter specially-acquired character.

It is not easy to distinguish between these later

acquired, or larval characters, and the features that are really due to inheritance; while the correct discrimination of them is one of the greatest problems which an embryologist has to solve. As a general rule, secondary characters will be more variable, because the different groups will probably have acquired them independently. Again, secondarily acquired characters must always be useful, must confer some advantage on their possessors, or otherwise they would never have been preserved; while on the other hand, structures such as rudimentary organs, which are of no practical use, must be inherited, for in no other way can their presence be explained.

One other cause of falsification of the ancestral history in actual development may be briefly alluded to. It happens not uncommonly that the larvæ and adults have entirely different habits, and in such cases the transition from one to the other is not always a gradual one, but may be effected by an abrupt, almost violent metamorphosis.

Take the case of a butterfly or moth. From the egg emerges a caterpillar, a soft-bodied vermiform animal with short fleshy legs adapted for crawling along the branches and leaves of plants, with jaws adapted for biting these leaves, and an alimentary canal fitted to digest them as food. The caterpillar feeds and grows rapidly, but retains its shape, and, except in size, makes no appreciable approach towards the adult condition. Having reached its full size, it changes into a chrysalis or pupa, and during this state, which lasts for weeks or months, it takes no food, having indeed no mouth, but lives at the expense of nutriment which it has accumulated in its body during the caterpillar stage. After a time, longer or shorter in different cases, the pupa skin is cast off and the full-blown butterfly or moth appears : an animal altogether different to the caterpillar ; provided with two pairs of wings, with three pairs of long jointed legs, with much more perfect sense-organs, and with its jaws modified so as to form a long tubular proboscis, by which it can suck up the juices of flowers on which it now feeds.

This is a typical case of abrupt metamorphosis, and it is obvious that the developmental history cannot here be a true recapitulation. It is quite impossible that the pupa, for instance, should ever have been an adult condition, and the abrupt character of the changes from caterpillar to pupa, and from pupa to imago, cannot be ancestral. Without entering at length into the origin of metamorphoses such as these, it may be pointed out that they only occur amongst insects, in forms in which the nature of the food, and therefore the structure of the jaws, is very different in the caterpillar and in the adult condition respectively ; and that in such cases a gradual transition from one to the other would be quite impossible, for a mouth intermediate in its characters between the masticatory mouth of the caterpillar and the suctorial one of the butterfly would manifestly be incapable of either biting leaves or sucking the juices of flowers.

In the case of other insects, such as the locust, cricket, or grasshopper, the developmental history is

one of gradual progression, and not one of abrupt
metamorphosis, each step being a step onwards
towards the adult insect. In the cockroach, again,
the process of development is gradual, and parts not
present in the larva, such as the wings, appear not
suddenly, but step by step and progressively. The
differences in these cases between the young and the
adult are not great.

There is no doubt that gradual transformation is
the simpler and more primitive condition, and that
the action of Natural Selection may cause the larval
and adult forms to move apart and constitute periods
of growth and reproduction respectively. Natural
Selection may also cause such specialisation of the
adult condition as to make it differ widely from the
larva in habits and the nature of its food as well as
in structure, and the larva and adult may move so far
apart as to render a period of quiescence necessary
in order to allow the change of organs into those
required for new work. Again, Natural Selection may
act on the larva as well, fitting it better for its own
life as distinct from that of the imago ; thus leading
to still further divergence, and resulting in the larva
acquiring characters that are no part of its ancestral
history.

A similar explanation applies to the process
whereby the young sea-urchin is formed within the
larva. The larva is adapted for a free-swimming
existence, the sea-urchin for crawling on the sea
bottom. A gradual transformation from one to the
other would be undesirable, for the intermediate conditions would be imperfectly adapted to either mode

of existence ; hence we find the external form of an
early ancestral stage preserved, while internally the
larva is passing through the later stages and gradually
working its way up to the adult form and structure.
A similar process occurs in many other animals from
various groups, the explanation in all cases being
found in considerations such as the above.

TESTS OF RECAPITULATION.

An important consideration is that, if the
developmental changes are to be interpreted as
a correct record of ancestral history, then, first,
the several stages must be all possible ones, the
history must be one that could actually have
occurred — i.e., the several steps of the history
as reconstructed must form a series, all the stages
of which are practicable ones.   Secondly, each stage
must be an advance of the preceding one, otherwise it would not have been retained, for it must
constitute an advance so distinct as to confer on its
possessor an appreciable advantage in the struggle
for existence.   It is not enough that the ultimate
stage should be more advantageous than the initial
ones, but each intermediate stage must also be a
distinct advance.   Intermediate stages, which are not
and could not be functional, can form no part of an
ancestral series.

A good example of an embryological series fulfil-
ling these conditions is afforded by the development
of the eye in the higher *Cephalopoda*.   First let us
consider the evolution of eyes in the Mollusca.   In

Solen we find the simplest condition of the molluscan
eye, merely a slightly depressed and slightly modified
patch of skin, which can only distinguish light from
darkness, and in which the sensitive cells are protected by being situated at the bottom of the fold of
skin.   In *Patella* the next stage is found, where the
eye forms a pit with a widely open mouth.   This is
a distinct advance on the preceding form, for owing
to the increased depth of the pit, the sensory cells
are less exposed to accidental injury.   The next stage
is found in *Haliotis*, and consists in the narrowing of
the mouth of the pit.   This is a simple change, but a
very important step forwards, for in consequence of
the smallness of the aperture, light from any one part
of an object can only fall on one particular part of the
pit or retina, and so an image, though a dim one, is
formed.   The next step consists in the formation of
a lens at the mouth of the pit, by a deposit of
cuticle : this form of eye is found in *Fissurella*.
(Fig. 17.)   The gain here is twofold—viz., increased
protection and increased brightness of the image, for
the lens will focus the rays of light more sharply on
the retina, and will allow a greater quantity of light,
a larger pencil of rays from each part of the object,
to reach the corresponding part of the retina.   (Fig.
18.)   Finally, the formation of the folds of skin
known as the iris and eyelids provides for the better
protection of the eye, and is a distinct advance on
the somewhat clumsy method of withdrawal seen in
the snail.   This is found in the Cephalopoda, such
as *Loligo*.

If now we study the actual development of the

Fig. 17.

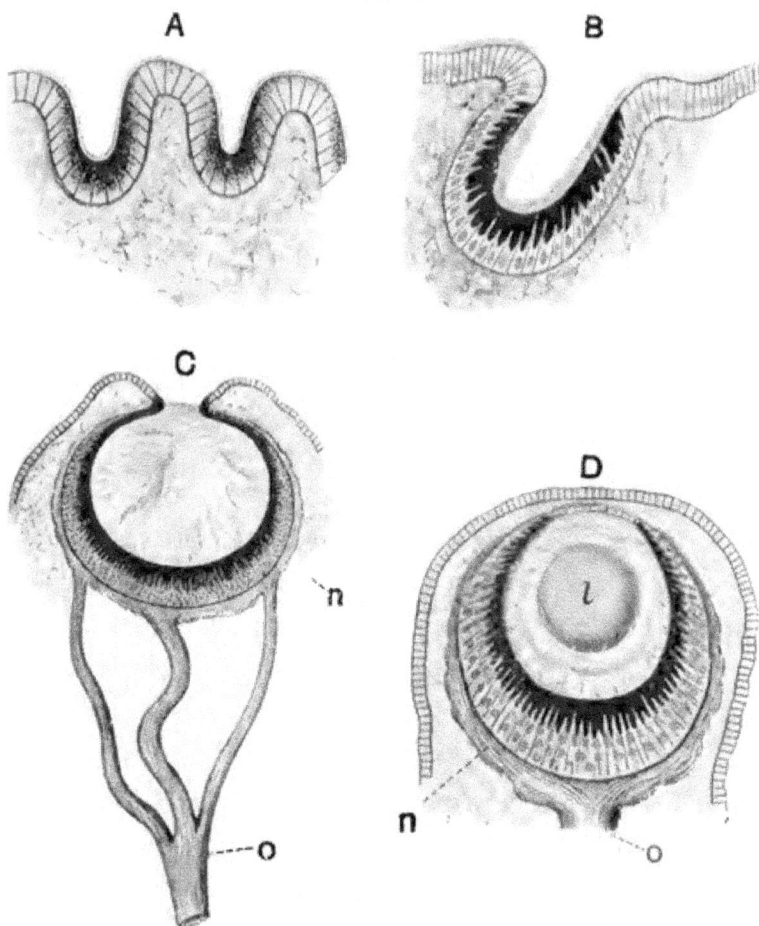

Evolution of eyes in Mollusca.

n, Layer of nerve tissue ; *o*, Optic nerve ; *l*, Lens.

A.—Eye of *Solen*, consisting of sensitive pigmented cells at the bottom of pit-like depressions in the skin.

B.—Eye of *Patella*. The pit is more developed, and the sensory cells less exposed.

C.—Eye of *Haliotis*. The mouth of the pit is narrow, and the sensitive cells still more protected. A well-developed optic nerve is present, and a layer of nerve tissue outside the sensitive cells of the retina.

D.—Eye of *Fissurella*. The mouth of the pit is closed, and a lens is developed.

eye of a cuttle-fish, we find that the eye, although a complicated one, yet passes in its own development through all the above series of stages, from the slight depression of skin, through the stages of a pit

Fig. 18.

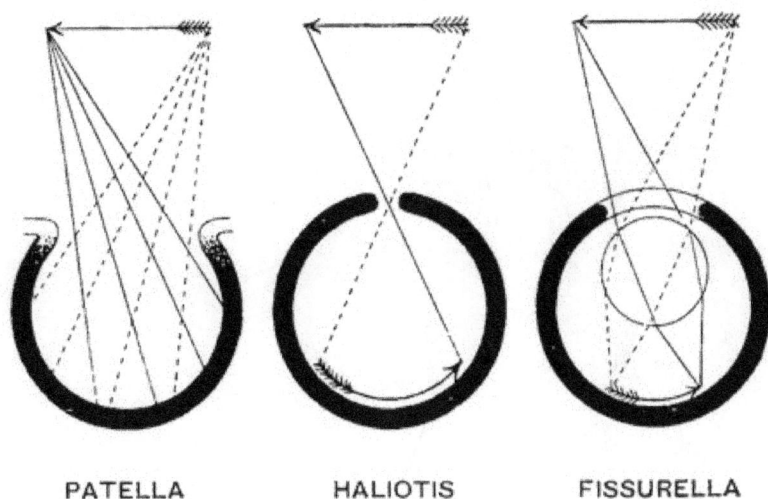

PATELLA   HALIOTIS   FISSURELLA

Diagram of eyes of Patella, Haliotis, *and* Fissurella.

In *Patella* no image is formed. In *Haliotis* an indistinct image is formed by narrowing of the mouth of the pit. In *Fissurella* a distinct image is formed by the lens.

with large and small mouth ; lens and finally eyelids being developed. (Fig. 19.)

The important point here is that we are able to show that the series fulfils our conditions, that all stages are possible ones, and that each is a distinct step onwards and an improvement on its predecessor ; and furthermore, that each stage is retained as the actual permanent condition in some actually living mollusc.

It is not always possible to point out so clearly as

FIG. 19.

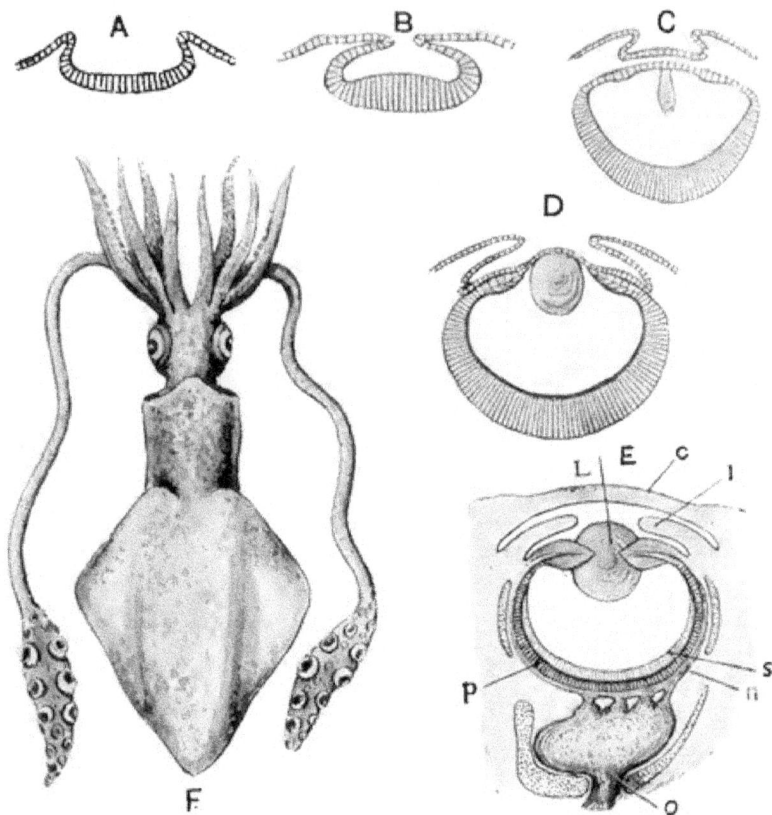

Loligo, one of the higher Cephalopoda, showing development of eye.

o, Optic nerve and ganglion ; *n*, Nerve layer of retina ; *l*, Lens ; *i*, Iris ;
c, Cornea ; *p*, Pigment layer of retina ; *s*, Layer of sensory cells.

A.—First stage ; a simple pit in the skin. This corresponds to the adult
condition in *Patella*.

B.—Narrowing of mouth of pit, corresponding to *Haliotis*.

C.—Closing of pit, and formation of lens, corresponding to *Fissurella*.

D.—Formation of iris.

E.—Adult condition.

F.—Figure of adult *Loligo*.

in the above instance the particular advantage gained
at each step, even when a complete developmental
series is known to us; but in such cases our difficulties may be largely ascribed to ignorance of the
particular conditions that confer advantage in the
struggle for existence.

Embryonic Stages viewed as Ancestors.

Early larval stages are of much interest, as
possibly indicating the forms of the earliest ancestors. The most important and fundamental point
is the fact that all the higher animals arise from
eggs, and that the bodies of the higher animals
are built up of cells or units, as a wall is built of
bricks.

The lowest animals, or *Protozoa*, are single units
or cells, and the egg of the higher animals is also a
single cell. Therefore each of the higher animals
begins its life as a single cell—*i.e.*, in a Protozoon
stage. Does not this indicate the descent of *Metazoa*
or multicellular animals from *Protozoa* or unicellular
ones? If there is a blood-relationship between the
highest and lowest animals, and if the higher are
descended from the lower, is it not reasonable to
look for the origin of Metazoa in the Protozoa?
May not this be the explanation of the origin of all
Metazoa in their actual development from single
cells, and may not the egg represent the Protozoon
stage in the ancestry?

Further, if animals really recapitulate, and if the
reason why all Metazoa begin life as single cells—

which is the most remarkable fact in the whole of embryology — is that they are descended from Protozoa, may not we hope to find from the study of early stages of development some hint as to the mode of origin of Metazoa in the first instance?

Let us consider, for example, the actual development of *Amphioxus*. The egg divides into a number of cells, which, instead of separating, remain together and continue to divide again and again, giving rise to the *morula* stage. The next stage is the *tubular* condition, where the cells are arranged regularly round a central cavity with an aperture at each end. This is followed by the *blastula* stage, which consists of a hollow ball, the outer cells of which are furnished with cilia enabling the embryo to swim freely. During later stages foldings take place, caused by outgrowths in some places and depressions in others, whereby the shape is gradually altered. (Fig. 20.)

Our present point is to ascertain whether these earliest stages are possible ones; whether there are organisms which remain permanently in one of these conditions—viz., (1) a single cell; (2) a heap of similar cells; (3) a hollow tube; (4) a ciliated hollow ball.

As examples of the first condition, or that of a single cell, the *Monads* may be taken. These are among the most minute and the simplest of living organisms, having an oval body, a nucleus, and a flagellum. They are found in infusions of animal and vegetable matter. An example of the second condition, a heap of similar cells, is found in *Pandorina*, a colony of similar flagellate cells, all alike and living together in a common capsule. The third condition is found in *Salinella*, one of the most recent discoveries, and one of the most remark-

FIG. 20.

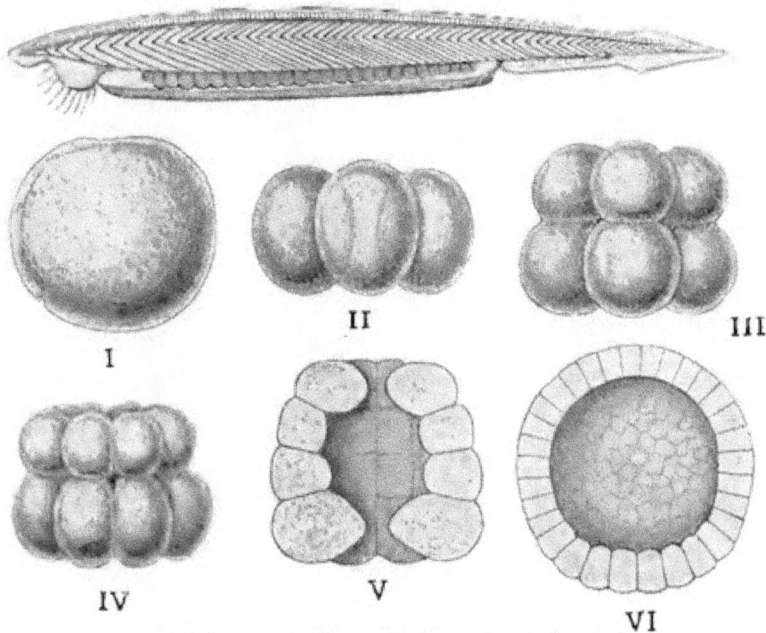

Amphioxus, *showing early stages in development.*

I. to IV.—Segmentation of the egg.
V.—Tubular stage, with an opening at each end.
VI.—*Blastula* stage, consisting of a hollow ball of cells.
The adult *Amphioxus* is shown at the top of the figure.

able animals known, which is found in water containing 2 per cent. of salt. This organism consists of a tube, open at both ends, the wall of which is formed of a single layer of cells. The fourth condition, that of a hollow ball or blastula, is represented by *Volvox*, the well-known fresh-water

organism, which consists of a hollow ball formed
by a single layer of flagellate cells like monads.
(Fig. 21.)

It must be noted that these examples do not form

FIG. 21.

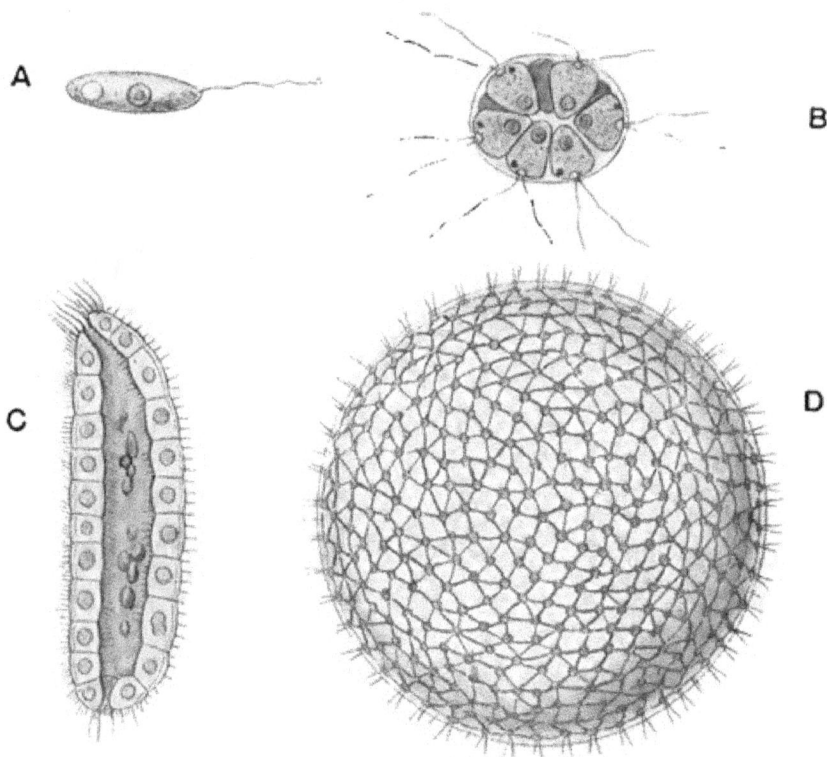

Adult organisms resembling early embryonic stages in higher animals.

A, *Monad;* B, *Pandorina;* C, *Salinella;* D, *Volvox.*

a continuous series in the sense of being derived
from one another. They are mentioned merely
with the object of showing that the very earliest
stages of development, like the later ones, may be
recapitulatory ; and that these early larval stages

represent possible forms of adult beings (whether animals or plants it is impossible to say) capable of independent existence ; and they illustrate very forcibly the interest which attaches to embryology in the light of the Recapitulation theory, coupled with the theory of Evolution with modification.

LECTURE V

THE COLOURS OF ANIMALS AND PLANTS

So universal and universally acknowledged are the power and charm of colour, and so accustomed are we to associate bright colours with health, happiness, and merriment, and gloomy colours with misfortune, evil, or with death, that we are apt tacitly or explicitly to assume that the existence of colour is sufficiently explained by the pleasure it gives us; that the exquisite and varied colours of butterflies, birds, and flowers, for instance, are developed and acquired for our special enjoyment. But this is not so, and the poet has warned us that this explanation is not enough, for "full many a flower is born to blush unseen, and waste its sweetness on the desert air." These lines man, in his supreme conceit, usually interprets as an expression of a half-contemptuous pity for the unfortunate flower which fails to meet his lordly eye, and so wastes alike its beauty and its sweetness. The poet Gray is, however, quite right, and there is something to be explained, as a moment's consideration will show us.

Birds and insects are the most gorgeously-coloured animals, and many of the most brightly-coloured are inhabitants of parts of South and Central America,

the Malay Archipelago, &c., where man is almost
unknown and wholly unwelcome. Again, man is a
comparatively recent arrival on the earth, and the
exquisite shapes and mouldings of many fossil shells
show us that at any rate beauty of form existed in
rare perfection long before his advent. This is a
subject worth inquiring into, and one which has
attracted the attention of many men, especially
Wallace. The result of these inquiries is to show
that colour is no mere accidental attribute of animals
and plants, but has a very definite reason for its
existence, and that in various ways and for divers
reasons it may contribute materially to the welfare
of its possessor.

The colours of animals in a state of nature are
constant, or nearly so. Tame rabbits vary greatly,
but wild ones are all very much alike ; each kind
of animal having a particular colouring, which does
not vary very greatly, save in exceptional cases.
Not merely in the individuals of a given species, but
in genera or even in entire families a certain con-
stancy of colouring may be noticed. For instance,
the "blue" family of butterflies are characterised,
not only by their blue colour, but more markedly
still by the eye-spots on the under-surface of the
wings. The mottled under-surface of the wings of
Vanessidæ, and the silvering of the wings of Fritil-
laries, are other examples.

Colour cannot be explained as due to the direct
action of light or heat ; for although it is true that
there is an immensely greater number of richly-
coloured birds and insects in tropical than in tem-

perate and cold countries, yet the majority of tropical birds are dull-coloured, and in some groups the most brightly-coloured members are not tropical : for instance, the Arctic ducks and divers are more handsome than the tropical ones. The humming-birds found in the Andes form another instance, for here they are confined to lofty mountains, some-times to a particular mountain, "just beneath the line of perpetual snow, at an elevation of some 16,000 feet, dwelling in a world of almost constant hail, sleet, and rain." Again, most tropical and brightly-coloured birds are denizens of the forests, and shaded from the direct action of the sun ; they abound also near the equator, where cloudy skies are very prevalent. In the case of flowers, Wallace remarks that "in proportion to the whole number of species of plants, those having gaily-coloured flowers are actually more abundant in the temperate zones than between the tropics."

NON-SIGNIFICANT COLOURS.

Many colours are the incidental results of chemical and physical structure, for the same reason that sulphate of copper is blue. The red colour of blood, the white colour of fat, the silvery colour of the bladder of many fish, the pigmented condition of the frog's peritoneum, and the green tint of the bones of many fish, are examples. The brilliant and varied colours of deep-sea animals are probably devoid of any significa-tion, and the green colour of grass and the blue colour of the sky, for all we know, are non-significant.

The red colour of *Tubifex*, for example, is associated with the physiological activity of hæmoglobin ; the red colour here is probably disadvantageous as such, but is counterbalanced by the physiological utility of the pigment for respiratory purposes. A similar explanation holds with chlorophyll, the green colouring matter of plants.

It is important to note that red is only red in the presence of light, and that a red animal if put in a dark place ceases to be red ; or if put in a green light, which it is incapable of reflecting. Non-significant colours "form the material out of which natural or sexual selection can form significant colours," and "all animal colours must have been originally non-significant."

The Direct Action of Environment.

Distinct colour varieties occur locally among the Lepidoptera, and a great prevalence of green is shown by the fauna of Ceylon, not only by terrestrial forms, but by echinoderms, corals, and other animals. That differences in food have an effect on colour has been shown by feeding the larvæ of various kinds of Lepidoptera on different plants: the larvæ of the eyed-hawk moth, the brimstone moth, and the peppered moth show changes of this kind. This effect is not due to the colour showing through the skin, but must be effected through the nervous system, the particular pigment being actually built up by the caterpillar.

Pupæ assume in many cases the colour of the

objects to which they are attached; for instance, it was shown that the pupa of the small tortoise-shell butterfly, when placed on a dark background, became itself very dark; but when placed on a white background it became light-coloured. This susceptibility to change of colour is greatest at the stage when the larva first fixes itself before changing to the pupa. The effect is produced through the skin generally, and not through the eyes.

Again, trout change colour according to that of the bottom of the stream they inhabit; so also do minnows; hence the interior of a minnow-can is painted white, so that the bait may be light-coloured, and more conspicuous to a pike or perch. Changes of colour depend on the eye in the case of fish, and a blind trout remains dark, the pigment cells relaxing and becoming flattened, thereby exposing their maximum amount of surface area. Cave animals, on the other hand, become pale, because the pigment which is now useless degenerates and disappears.

The effect of cold in causing change of colour indirectly through the nervous system, has been demonstrated by suddenly exposing animals to cold which had previously been protected from it for some time, the result being distinct blanching. Melanism, or dark coloration, is common on oceanic islands, and humidity of the atmosphere is as a rule associated with the darkening of colours. So it is with increased elevation, and it is possible that the object of this is to increase the absorption of heat. Brilliant colours are not dependent on or proportionate to the amount of light.

The effect of environment in causing changes of colour is well shown in cases of what is known as *seasonal dimorphism*, where animals produce two broods in each year, each of different appearance with regard to colouring, and each capable of producing the other.   A good example of this is found in the two continental butterflies *Vanessa prorsa* and *Vanessa levana.*

Vanessa levana, the spring form, has a red ground colour, with black spots and dashes, and a row of blue spots round the margin of the hind wings ; *Vanessa prorsa,* the summer form, is deep black, with a broad yellowish-white band across both wings, and with no blue spots.   These were formerly called distinct species, but have recently been shown to be varieties of one and the same species.   That these differences are due to the direct action of cold and heat has been shown by keeping the pupæ of *levana* at a low temperature.   These, which would ordinarily have produced the summer form *prorsa,* hatched, under the altered conditions of temperature, partly as *levana* and partly as an intermediate form.

Again, very young Canaries change their colour to orange when given Cayenne pepper, and certain parrots have been shown to change their colour when fed on the fat of a particular fish.

It is very difficult to draw the line between the direct action of environment through the nervous system, and the action of Natural Selection ; for to which can we attribute the whitening of Arctic animals ?

We have now to consider the great mass of cases

illustrating the preservation and accentuation of colour through the agency of Natural Selection— one of the most striking of the later developments of the theory.

Significant Colours.

These are colours which are of direct advantage to their possessor as colour, and not merely because they are associated with other properties which are useful, as in the case of hæmoglobin and chlorophyll. The classification of these colours is a matter of some difficulty, for cross-relations occur which are difficult to express.

There are three chief classes or groups :—

1. *Apatetic;* the purpose of which, or rather the object gained by which, is to hinder recognition by other animals.

2. *Sematic,* or signalling colours; the purpose of which is to facilitate or aid recognition by animals of the same or of other kinds.

3. *Epigamic;* which include those cases in which differences occur between the male and female sex, as in the peacock and pea-hen, the duck and drake, &c. This is a special and important group.

Apatetic Colours are again divided into:

    a. Protective resemblances, aiding escape from enemies, as in those cases where animals resemble sticks or plants, and so escape notice.

    b. Aggressive resemblances; the purpose of which is to aid the approach to prey; for example the resemblance of the colour of the lion to that of the desert.

  c. Alluring resemblances; constituting a small
   group of cases, in which an animal acts as a
   bait by taking on the form of something
   attractive to its prey.

SEMATIC COLOURS have two subdivisions :

  a. Warning colours. These constitute a curious
   group of cases, in which animals have bright
   conspicuous colours, for the purpose of
   warning other animals off them, and which
   are signs of inedibility or of the possession
   of dangerous powers of attack.

  b. Recognition colours. These are for the purpose
   of easy recognition by animals of the same
   kind ; and are best seen in the cases of
   gregarious animals, such as deer, whose
   safety largely depends on association and
   mutual defence.

PROTECTIVE RESEMBLANCES.

Such forms of protective colouring as aid the
escape from enemies by hindering recognition may
be one of two kinds :

(1) *General;* in which the colouring is such as to
assimilate the animal to its environment, and so
render it less conspicuous, as in the case of the
whiteness of Arctic animals, such as the Polar bear ;
and the sandy colour of desert animals, or the transparent blueness of pelagic forms.

(2) *Special;* where the resemblance is to some
particular object, and where the animal escapes, not
through being concealed from view and so over-

looked, but through being mistaken for something else. Of these cases some extraordinary instances are known. The resemblance may be to another animal or to a plant, flower, or leaf, or to inorganic substances.

Furthermore, the protective colouring may be either constant or variable; a good example of variable protective colouring being shown by the *Octopus* and *Chameleon*. Again, in animals such as insects, which undergo metamorphosis, and in which the form, structure, and habits are widely different in the larval and adult stages respectively, both these stages may be protectively coloured, but the resemblance will be to entirely different objects.

Let us take examples from the different groups of animals, and we shall see that the reality of protective colouring is impossible to doubt.

MAMMALS.—The whiteness of Arctic animals has already been referred to. The American polar bear is white all the year round; the ermine or stoat changes to white in the winter, and the Arctic fox usually does this also. The Alpine hare always becomes white in the winter in Scandinavia, and usually in Scotland, although rarely so in Ireland. This change consists in an actual blanching of the hairs from the tips inwards, with a new growth of additional white hairs. The general tawny colour of deer is also protective; the protection afforded by spots is seen by their resemblance to the circular spots of light caused by sunlight passing through the leaves of a wood, while stripes facilitate escape in long grass or reeds.

To fully appreciate the protective value of colours it is necessary to see the animals in their native haunts. Thus, speaking of the Zebra, Francis Galton says that although "no more conspicuous animal can well be conceived," yet the proportion of the black and white stripes "is such as exactly to match the pale tint which arid ground possesses when seen by moonlight." With regard to the Giraffe, Sir S. Baker graphically describes the way in which, when seen at a distance, it resembles a dead tree stem. Again, the green colour of the Sloth is due to parasitic algæ, which cause it to resemble a lichen-covered branch; an oval buff-coloured mark on the back giving the impression of the broken end of the stump.

BIRDS.—The summer plumage of the Ptarmigan conceals it among the heather very effectively; in winter it becomes white. The Heron again, is almost impossible to find among the rushes, where it stands in an absolutely vertical position, with the tip of its beak tilted up. The ventral surface having pale yellow stripes, closely resembles the surrounding rushes. Moreover, it turns slowly round so as always to present the ventral surface to view, while the striped back and broad dark-coloured sides are never presented to the observer. In the case of birds which build open nests, the female is protectively coloured : this is well seen in the pheasant. When both sexes are brilliantly coloured, such as the king-fisher, parrots, &c., the nest is of such a nature as to conceal the sitting bird.

INSECTS.—These afford the best examples of protective colouring both in the larval and adult states.

1. *Larval Insects.*—Lepidopteran larvæ, or caterpillars, are in the great majority of cases absolutely
defenceless : their bodies are soft and their very
shape is due to their containing fluid under pressure,
therefore a slight wound involves much loss of fluid
or blood ; hence their great need for protection.
The great purpose in life of a caterpillar, next to
feeding, is not to be seen, or rather not to be recognised.

General protective resemblances are found in their
green colour, which is the most usual, and harmonises
with that of the food-plant. This green colour is
partly due directly to food and partly to metachloro-
phyll, a special pigment in the blood, and a slightly
altered derivative of the chlorophyll of the food.
Those caterpillars which have the habit of feeding
either on grass, or on low-growing plants among the
grass, are protected by longitudinal striping. In the
larger caterpillars, such as those of the Privet hawk
moth, which are striped transversely or obliquely,
the colour is usually that of the flower of the food-
plant, and the stripes serve to break up the surface
of the body. These larvæ turn brown at the time of
descending to the earth to change into pupæ.

Special protective resemblances are best seen in the
larvæ of the geometer moths, or "stick caterpillars"
as they are called. These are very common, and are
rarely seen, or rather detected, owing to their resemblance to twigs. They have only two pairs of
legs or claspers, a long thin and cylindrical body,
which stands out at an acute angle with the stem
upon which they fix themselves, and upon which

they sit motionless for hours ; this absence of movement being very important in order to increase the
deception. The head is modified in shape to increase the resemblance to twigs, and a silk thread is
spun, attached to the twig, to relieve the tension
involved by remaining in the same position for so
long a time. They feed at night, when there is less
need for protective devices. The protection here is
so real that a green lizard will generally fail to detect
a stick caterpillar in its position of rest, though it
will seize and greedily devour it directly it moves.

2. *Adult Insects.*—In butterflies the under surface
of the wings is coloured protectively, the upper surface
attractively ; and the sudden change when they fold
their wings over their backs is often enough to defy
detection. The most perfect examples are found in
the leaf butterflies, such as *Kallima*, found in the
Malay Archipelago, India, and Africa. This is a
very common and showy butterfly, with orange and
purple colouring on the upper surface of the wings.
It is a rapid flier, and frequents dry forests, always
settling where there is dead and decaying foliage.
The colouring on the under surface of the wings
bears a remarkable resemblance to that of a dead
leaf, and when the wings are turned up, with the head
and body hidden between them, it is often very
difficult to distinguish it from dead leaves, the
resemblance being rendered even more close by the
short tail, which looks like the stalk of a leaf, and by
the markings on the under surface, which closely
imitate the mid-rib and veins of a leaf. Speaking of
this insect, Mr. Wallace says : " The colour is very

remarkable for its extreme amount of variability, from deep reddish-brown to olive or pale yellow, hardly two specimens being exactly alike, but all coming within the range of leaves in various stages of decay. Still more curious is the fact that the paler wings, which imitate leaves most decayed, are usually covered with small black dots, exactly resembling the minute fungi on decaying leaves." (Fig. 22.) This is an extreme case of what is really a general law among butterflies. The mode in which it is acquired is as follows : At first there is a more or less accidental resemblance. Large numbers of butterflies are killed by birds, lizards, and other animals, and any whose markings and habits of perching render them less easy to detect will have a better chance of escaping, and so of laying eggs and transmitting their peculiarities to their offspring. This protection becomes, through selection, better from generation to generation, and the imperfectly protected forms are weeded out and eaten.

FIG. 22.

Kalima, *showing under surface of wing.*

Similar examples are found in the Herald and Angle shade moths, which resemble decayed and crumpled leaves; in the Buff-tip moth, which re-

FIG. 23.

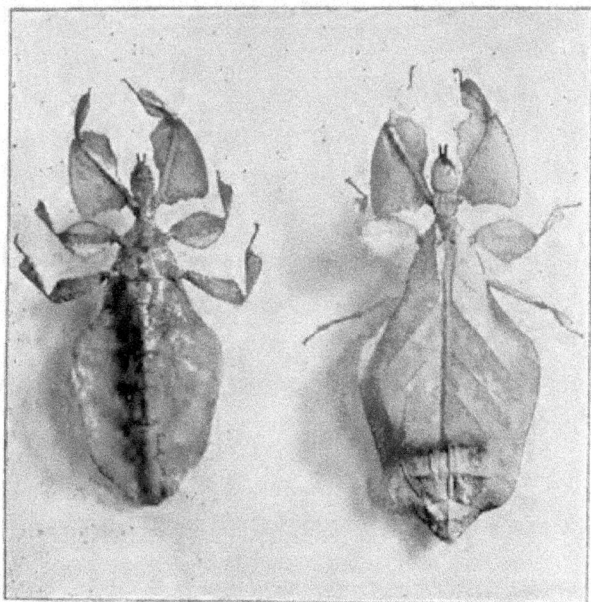

Phyllium, *the Leaf Insect.*

FIG. 24.

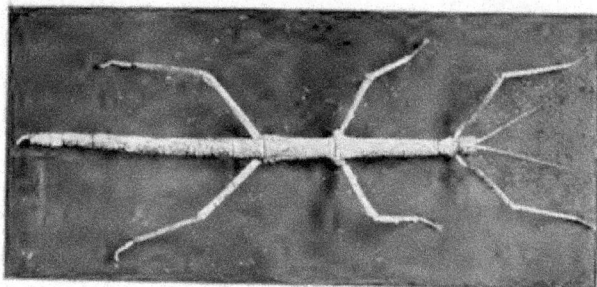

Lonchodes, *the Stick Insect.*

sembles a broken piece of decayed and lichen-
covered stick; and in the Lappet moth which has the

I

appearance of dried leaves. But the best example perhaps is that of *Phyllium*, the leaf insect found in the East Indies, which has in the adult condition a most extraordinary resemblance to a leaf, or rather a bunch of leaves. The colour is green, and the wing-cases are marked with lines like the veins of a leaf : some of the joints of the limbs are flattened and expanded. Andrew Murray relates how an Indian species exhibited in the Botanical Garden at Edinburgh deceived everybody by its resemblance to the plant upon which it lived. The deception was ultimately the cause of its death ; for the visitors, sceptical as to its animal nature, insisted on touching it before they would be convinced. (Fig. 23.)

Variable Protective Resemblances.

Instances of this we have already mentioned in the case of the Trout, which changes colour according to that of the bottom of the stream. This effect is always produced through the eye, and is consequently absent in blind animals. How far it is under the control of the will is difficult to determine. Another excellent illustration is afforded by the Chameleon and many lizards which possess the power of changing colour. This power depends on the presence of chromato-phores or pigmented bodies found in the skin, which have the power of changing shape. These are of different colours, and are arranged in layers, some near the surface, some deeper ; the light yellow cells being most superficial, then the red and brown, and the black deepest of all. If the superficial ones contract, the deeper ones will become more apparent, and *vice versâ*. These changes of colour always bear a relation to the surface on which the animal is placed at the time, and are therefore supposed with good reason to be protective.

AGGRESSIVE RESEMBLANCES.

These aid the approach of their possessor to its prey, through a superficial resemblance to its general environment or to other objects. The Lion and Tiger are good examples of general aggressive resemblances, the lion resembling in colour the desert in which it lives, and the stripes on the tiger rendering it easy to conceal itself among the long grass. So with the Polar bear, which not only is of the same colour as the snow, but also has a curious shapeless outline when lying down, not unlike a heap of snow.

ALLURING RESEMBLANCES.

These attract or entice the approach of prey. *Lophius piscatorius*, the angler-fish, which conceals itself in the mud at the bottom of the water, is provided with long tentacles, which are mistaken by small fish for worms writhing in the muddy water. The body of the angler, being concealed in the mud, is not seen, and the prey, deceived by what they imagine to be worms, are themselves eaten by the angler-fish. Again, certain deep-sea fish have a luminous phosphorescent organ at the end of the foremost tentacle, suspended as a lure in front of the mouth, to attract their prey.

One of the lizards, *Phrynocephalus mystaceus,* found in Asia, not only in its general colour resembles sand, but is also furnished with a red fold of skin at each angle of the mouth, which is produced into a flower-like shape resembling a little red flower which grows in the sand. Insects, attracted by what they believe to be flowers, approach the mouth of the lizard, and fall victims to its snare. The Indian Mantis, *Hymenopus bicornis,* feeds on other insects, which it attracts by its flower-like shape and pink colour, which resemble an orchid, the shape being due to flattening of the proximal joints of the limbs, which radiate from the body like the petals of a flower. *Thomisus decipiens,* a spider found by Forbes, and seen twice only, resembled exactly the droppings of birds on a leaf.

ADVENTITIOUS COLOURING.

This may possibly be either aggressive or protective. The Caddis worms, for instance, make cases of sand, shells, twigs, and any objects found at the bottom of streams. *Xenophora,* a genus of Gasteropods, build pieces of dead shells, rocks, and corals into the edge of their growing shells. Some crabs, again (*Stenorhynchus*), have a habit of fastening pieces of sea-weed, &c., on to their bodies or limbs. Mr. Bateson describes how these crabs tear the weed in pieces with their claws, and after chewing the pieces in their mouths in order to soften them, rub them on their head or legs till they are caught by the hairs which cover them. "The whole proceeding is most human and purposeful." These processes are gone through both by night and day, and a blinded Stenorhynchus, if cleaned of its weed, "will immediately begin to clothe itself again with the same care and precision as before." It shows no disposition to take up a position among similarly coloured objects ; and some individuals which have taken up stations among weeds do not dress themselves at all.

WARNING COLOURS.

If an animal belonging to a group liable to be eaten by others, is possessed of a nauseous taste, or if an animal, such as a wasp, is specially armed and venomous, it is to its advantage that it should be recognised quickly, and so avoided by animals that might be disposed to take it as food.

Hence arises *warning coloration*, the explanation of which is due to Wallace. Darwin, who was unable to explain the reason for the gaudy coloration of some caterpillars, applied to Wallace, stated his difficulty, and asked for suggestions. Wallace thought the matter over, considered all known cases, and then ventured to predict that birds and other enemies would be found to refuse such caterpillars if offered to them. This explanation, first applied to caterpillars, soon extended to adult forms, not only of insects, but of other groups as well.

An excellent example is afforded by the Skunk (*Mephitis mephitica*), a small black and white animal possessing a very offensive secretion, which it ejects over its enemies, and which protects it from their attacks. Owing to this offensive weapon the skunk is seldom attacked by other animals, and its black and white coloration easily distinguishes it from unprotected animals.

Insects afford many admirable examples of warning colours, and many well-known instances are found among butterflies. The best examples among these are found in three great families of butterflies—the *Heliconidæ*, found in South America, the *Danaidæ*, found in Asia and tropical regions generally, and the *Acræidæ* of Africa. These have large but rather weak wings, and fly slowly. They are always very abundant, all have conspicuous colours or markings, and often a peculiar form of flight; characters by which they can be recognised at a glance. The colours are nearly always the same on both upper and under surfaces of the wings; they never try to conceal themselves, but rest on the upper surfaces of leaves and flowers. Moreover, they all have juices which exhale a powerful scent; so that if they are killed by pinching the body, a liquid exudes which stains the fingers yellow, and leaves an odour which can only be removed by repeated washing. This odour is not very offensive to man, but has been shown by experiment to be so to birds and other insect-eating animals.

Warning colours are advertisements, often highly-coloured advertisements, of unsuitability as food.

Insects are of two kinds—those which are extremely
difficult to find, and those which are rendered
prominent through startling colours and conspicuous
attitudes.   Warning colours may usually be distinguished by being conspicuously exposed when the
animal is at rest.   Crude patterns and startling
contrasts in colour are characteristically warning, and
these colours and patterns often resemble each other;
black combined with white, yellow, or red, are the
commonest combinations, and the patterns usually
consist of rings, stripes, or spots.

Other examples are found in the bright colours of
some sponges which have been proved to be nauseous
to fish ; in the Anemones, Ascidians, and many
brightly-coloured Nudibranchs.

One of the best-known instances is that of the
frog found by Mr. Belt in Nicaragua, a small
animal, gorgeously coloured with red and blue, which
never hides itself; whereas most frogs are coloured
green or brown, and hide during the day-time,
to avoid being eaten by snakes and birds.   Sus-
pecting this animal to be uneatable, Mr. Belt
offered it to ducks and fowl, all of which refused to
touch it, except one young duck, which took the frog
in its mouth, but dropped it directly, "and went
about jerking its head, as though trying to throw off
some unpleasant taste."

Very numerous examples are found among caterpillars such as that of the Cinnabar moth, which
is coloured black and yellow, and rejected even by a
toad.   The Magpie-moth caterpillar, which is cream-
coloured, with orange and black markings, and

extremely conspicuous, is either refused altogether by birds, lizards, frogs, and spiders, or else causes them to exhibit signs of most intense disgust after eating it. The caterpillars of the Tiger-moth, the Burnet, and the Buff-tip are all brightly coloured and nauseous.

It has been objected, first, that the protection afforded by warning colours is only imperfect, that it must not be overdone, and is only available to a few ; that the likes and dislikes of insect-eating animals are purely relative, and hunger will drive them to extremes and overcome taste, hence giving rise to contradictory results of experiments. But this is the very essence of natural selection, which preserves advantageous characters, but does not lead to perfection. Secondly, it has been stated that "tasting is quite as dangerous to the caterpillar as swallowing outright," and hence it is argued that there is no advantage. But this, again, is a misconception, for although the individual may suffer, the species will benefit through the lesson learnt by its death.

Eisig's theory of warning colours states that the pigment itself is the cause of the distastefulness, and that it is very probably excretory in nature. According to this, the brilliant colours—*i.e.*, the abundant secretion of pigment—have caused the inedibility of the species, rather than that the inedibility has necessitated the production of bright colours as an advertisement. Brilliant colouring is the normal condition in caterpillars, and the advent of birds led to protective modifications, except when combined with inedibility.

MIMICRY.

Examples of mimicry are really cases of protective colouring, but are only intelligible through knowledge and appreciation of the value of the warning colour. Alluring colours, such as those of the *Mantis*, which simulates an orchid, may aptly be described as cases of " wolves in sheep's clothing," while the cases of mimicry may be considered as "asses in lions' skins."

Warning colours are conspicuous advertisements of inedibility, and certain colours and groupings of colours are usual, so that the lesson may be more easily learnt.   Black, white, red, and yellow, in startling and striking contrasts, form the usual types of warning colours.   If these are successful—*i.e.*, generally recognised as signs of inedibility—it is clear that other and different animals, which resemble them sufficiently closely to be mistaken for them, might benefit by the mistake, and escape.

Certain butterflies, the best examples of which are found among the *Heliconidæ, Danaidæ*, and *Acræidæ*, are nauseous, slow-flying, gaudily-coloured insects, having an unpleasant smell, and taking no pains to conceal themselves. Alongside these occur edible forms belonging to totally different genera and families, each of which shows a striking resemblance to one particular species of the protected butterfly ; this being in many cases confined to the female, which has greater need for protection.   A large number of cases of this mimicry are now known ; for instance, *Leptalis*, a form allied to the common garden white, mimics *Methona*, one of the protected Heliconidæ, in the shape of its body and wings, in its colour and even in its habits and mode of flight; so much so

FIG. 25.

Methona (protected).

Leptalis (mimetic).

that they are difficult to distinguish from one another. (Fig. 25.)

Other examples of mimicry are found in the Beehawk moth, which mimics the humble-bee; in the

Clearwing moths and Hornet moths, in which protection is obtained by mimicking insects possessing stings, and some of these are even said to writhe the abdomen when captured, as though pretending to sting; others are said to have the characteristic odour of the hornet. The "devil's coach-horse," the beetle with the habit of turning its tail over its back, pretending to have a sting, certainly deceives children, and perhaps grown-up people as well.

There is no doubt as to the success of this deception, which may deceive expert naturalists, or even the insects themselves; and Fritz Müller says: "I have repeatedly seen the male pursuing the mimicked species, till after closely approaching and becoming aware of his error, he suddenly returned."

Perhaps the most remarkable instance known is that of *Papilio merope*, one of the South African swallow-tail butterflies. The female of this species alone mimics; and of the female three forms are known, each of which mimics a different species of *Danais* prevalent in its own district. In Madagascar, which in so many other instances furnishes us with a glimpse of what the ancestral African fauna must have been, the female *Papilio merope* closely resembles the male, and is not mimetic.

In butterflies the female is more in need of protection, because it is of slower flight and liable to be exposed to attack while laying eggs. All an adult butterfly has to do is to pair and lay eggs; and as these events take up a very short time, protective colouring is not very useful, but warning colours are far more so.

Among birds, the Cuckoo is undoubtedly protected by its similarity to a hawk in appearance and in mode of flight. In the snakes, again, the harmless ones are said to mimic the venomous.

The conditions necessary in order to effect mimicry are given by Wallace as follows :—

(1) The two species, the imitating and the imitated, must occur in the same locality.

(2) The imitating species must be the more defenceless.

(3) The imitating species must be less numerous than the imitated, in individuals.

(4) The imitating species must differ from the bulk of its allies.

(5) The imitation, however minute, is external only, never extending to internal characters or to such as do not affect external appearance.

The mode of acquisition of mimicry is by the gradual action of natural selection, and must have been accidental in the first instance.

Recognition Markings.

These are closely allied to warning colours, and their purpose is to facilitate recognition, not by enemies, but by friends. They are seen especially in gregarious animals, and specific markings and colours are very probably in many cases not protective, but for recognition. A good instance of this class of colouring is seen in the upturned white tail of the Rabbit, which, although making it conspicuous to its enemies as well as friends, is probably a signal of danger to the other rabbits; and when feeding together, in accordance with their social habits, soon after sunset or on moonlight nights, the upturned tails of those in front serve as guides to those behind to run home on the appearance of an enemy. Many birds, antelopes, and other animals, have markings believed to serve a similar purpose, and probably the principle of distinctive colouring for recognition has something to do with the great diversity of colour met with in butterflies.

EPIGAMIC COLORATION.

This is seen in mature animals, especially in butterflies and birds, where the two sexes differ markedly as regards colour. As a general rule, the male is of the same hue as the female, but of a deeper and more intensified colour; for instance, in thrushes, hawks, and in the Emperor moth. Sometimes patches of colour found in the males are absent in the females, as in the Orange-tip butterfly. In some cases there are more extreme differences, as in the drake, peacock, cocks and hens, pheasants and Bird of Paradise; gay colours being the special privilege of the male. (Fig. 26.)

It is curious to note how with man the conditions are reversed, for the female butterfly or bird is as a rule larger and plainer than her mate. So it is with the organs of voice; the male cricket or grasshopper can alone produce sound, and many female birds have no song. The power of talking was originally the exclusive possession of the males—a privilege that with us they have long had to surrender.

The origin of the bright colours of the male may possibly be due to the selective action of the sexes

FIG. 26.

Bird of Paradise, male and female.

on one another. As among deer now, and men in several countries, the best fighters carry off the prizes; so among birds and butterflies the most gaily or tastefully arrayed males are the most highly favoured, and have the best chance of securing the most eligible mates. In the case of the Argus pheasant, the wing feathers are enormously elongated and of marvellously beautiful colouring; while during courtship the male erects his tail like a fan, displaying his glories to their best advantage. Similar examples are afforded by humming-birds.

There is no possible doubt as to the appreciation of colour by animals, and the chief differences between men and butterflies would appear to lie in the infinitely better taste displayed by the latter in the selection and combination of colour, both as regards marvellously delicate gradations of tint, and as regards daring combinations of strongly-contrasted colours. It is in fact as rare for a bird or butterfly to offend against good taste in matters of colour, as it is for a man to conform to it.

The theory of sexual selection, which was proposed by Darwin as supplementary to natural selection, is disputed by Wallace, who holds that brighter colour is the physical equivalent of greater vigour.

THE COLOURS OF FLOWERS AND FRUIT.

The essential parts of a flower are the ovary, in which the ovules are produced, and the anthers, in which the pollen is contained; and in order that the ovule may give rise to a seed—*i.e.*, to something capable of growing into a new plant—it must be fertilised by the pollen. There is a great advantage as regards the number of seeds produced, and the vigour of the offspring, if the ovules are fertilised by pollen, not from the same flower, but from a different flower or plant. This *cross fertilisation* is always highly beneficial and often absolutely essential. It is effected mainly by the agency of insects, especially bees, flies, and butterflies. These are induced to visit the flowers by bribes of honey secreted by the flower in such a position that, in order to reach it, the insect must brush against the anthers and get dusted with the pollen, by which, on visiting a second flower, fertilisation is effected.

The purpose of the coloured part of the flower is to form a conspicuous advertisement to insects of places where honey is to be found; and more detailed markings in the flower direct the insect towards the store of honey. Curiously ingenious contrivances are found in order to prevent self-fertilisation, and to ensure that the insect shall effect its work properly.

A familiar instance is that of *Orchis mascula*, the spotted orchid, which is abundant in meadows and in damp places in open woods. This consists of a spike of flowers, the calyx of which is formed by three coloured sepals, and the corolla by three petals. One of the petals, called the *labellum*, is larger than the others, and forms a sort of landing-stage. This is prolonged backwards into a spur-like nectary with spongy walls. The male organs consist of one anther with two cells, each of which contains a pollen mass. The ovary has three pistils, united together and twisted, ending above in two almost confluent stigmas. The third stigma forms the *rostellum*, a rounded projection overhanging the other stigmas and the entrance to the nectary. The pollinia, which lie in the anther cells, are club-shaped, the head of the club consisting of a number of packets of pollen grains united by thin elastic threads : the stalk of the club ends in a disc with a ball of very viscid matter on its under side, lying in the rostellum. The anther cells open when ripe, exposing the pollinia ; the rostellum is very delicate, and is ruptured by the slightest touch, exposing the viscid balls.

The problem is to transfer the pollinia from one spike of flowers to another. The manner in which this problem is solved through the agency of insects is as follows.

The insect, alighting on the labellum, pushes its head into the flower in order to reach the spur with its proboscis. In doing this it knocks against the rostellum, displacing its covering membrane and exposing the viscid balls to which the pollen masses are attached. On withdrawing its head the pollen masses come away firmly cemented to it and standing erect. In about thirty seconds the viscid disc contracts, causing the pollen mass to bend forwards through an angle of 90°, so as to become horizontal. By this contraction the pollen mass will be in a position to be applied directly to the stigma, when the insect visits the next flower. This manœuvre can be imitated by pushing the point of a pencil into a flower as shown in the figure, when the pollen masses will come away fixed to the pencil. (Fig. 27.)

In this way an insect flying from flower to flower

K

FIG. 27.

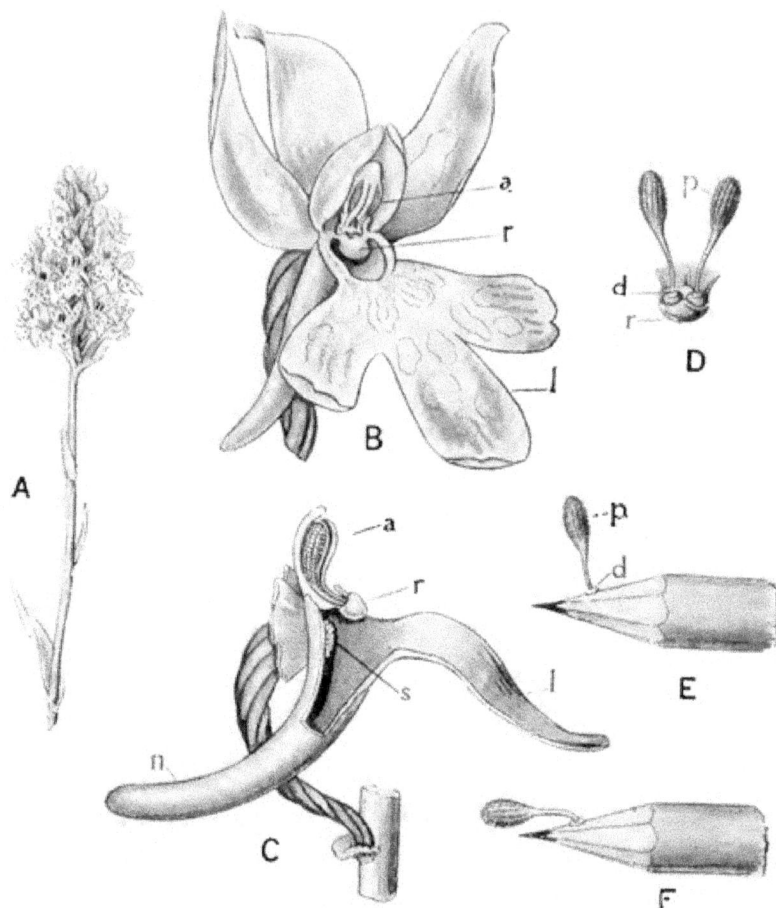

Orchis mascula.

a, Anther; *r*, Rostellum; *l*, Labellum; *p*, Pollen-mass; *d*, Disc at base of pollen-mass; *s*, Stigma; *n*, Nectary.

A.—Spike of flowers.
B.—Single flower.
C.—Flower dissected to show relations of anther, rostellum and stigma.
D.—Front view of pollen masses, with their discs lying in rostellum.
E.—Pollen-mass when first attached.
F.—Pollen-mass depressed, ready to effect cross-fertilisation.

effects cross-fertilisation regularly, and humble bees were actually watched in the act of fertilising by Hermann Müller. He saw them insert their heads into the flower and emerge with the pollinia attached, visit other flowers on the same spike, where they tried, more or less ineffectually, to rub off the pollinia, and finally fly off to other plants. Out of 97 humble bees which he caught, 32 bore the pollen masses of orchids. He proved that the bees visit the flower to obtain the fluid in the nectary, the walls of which they pierce with their maxillæ. Moreover, he timed the bees and found that they spent three or four seconds at each flower ; two or three seconds being sufficient to fix the pollinia. The average time spent at a given spike of flowers was twenty to twenty-two seconds, the bees then flying to another spike. In twenty-five to thirty seconds the pollinia were depressed and cross-fertilisation ensured.

The beauty and odour of flowers and the storage of honey are thus due to the existence of insects, and in a large number of cases the actual insects are known which effect cross fertilisation. Such is the case with regard to all conspicuous flowers : honey is secreted in order to attract insects, and the flowers are large and conspicuously coloured, so as to be readily seen by them. A striking illustration of this is seen in the common Clover. Darwin showed that by protecting 100 flowers with a net, not a single seed was produced from them ; whereas the 100 flowers which were outside the net were visited by bees and produced 2720 seeds. Hence, but for humble bees, which are the only insects visiting the common red clover, there would soon be no clover.

Large conspicuous flowers are visited much more frequently and by many more kinds of insects than are small inconspicuous ones. The long tubular corolla of many flowers is acquired so that certain insects alone should be able to get at the honey, these insects being the ones best suited for fertilising the flower.

The bright colour of the whole flower is to attract insects at a distance; the coloured dots and lines on the petals serve to guide it to the store of honey. This fact was proved by Darwin, who cut off the petals of *Lobelia*, and found that these flowers were then neglected by bees, which were perpetually visiting the other flowers.

In the case of flowers which are fertilised by means of the wind, such as grasses and trees, the flowers are small and not gaily coloured, and possess an enormous amount of pollen and a very large stigma. Moreover, in localities where insects are few in number, we find the flowers very insignificant in colour: for example, in the Galapagos Islands, which have only one butterfly and no bees.

White flowers are fertilised by nocturnal insects, chiefly moths. These flowers are always odorous, the jasmine and clematis for example, and often odorous only at night. Alpine flowers, again, are peculiarly beautiful, and the size of individual flowers is increased owing to the comparative scarcity of insects in the places where they grow, and the consequent necessity of attracting them from afar.

Fruits.

Fruits consist of seeds with surrounding envelopes of various kinds, and require to disperse their seeds so as to reach places favourable for growth and germination. Dispersion of the seeds is effected in some cases, such as the dandelion, by means of the wind ; in the edible fruits, on the other hand, it is effected by the fruit being swallowed by animals as food. Fruits are divided into two great groups— *attractive fruits* and *protective fruits*.

ATTRACTIVE FRUITS are soft, pulpy, and agreeable to the taste—such as the cherry, grape, strawberry, &c.—and are devoured by birds or mammals. In these the seeds themselves are hard, and pass through the animal unchanged. It is probable that every brightly-coloured pulpy fruit serves as food for some species of bird or mammal.

PROTECTIVE FRUITS, such as nuts. In these the part that would be eaten by an animal is the seed itself, and this is protectively coloured, being green while on the tree, and turning brown as it ripens and falls to the ground. Many seeds are specially protected, such as the chestnut by its prickly coat, and the walnut by its nauseous covering.

It thus appears that we owe the existence of many flowers to insects, and of many fruits to birds and mammals. This is an excellent example of the interest imparted to everyday life by the theory of Natural Selection, which tells us that we have merely to watch closely, to note carefully what is going on every day before our eyes, in order to obtain the

clue to problems of extraordinary and widely spread
interest. On the other hand, the support this theory
receives through being able to offer a ready and
complete explanation of so many and such divers
facts is very great indeed, and becomes all the more
significant when we reflect that the facts themselves
only came to light, or received serious attention,
some time after the promulgation of the theory.

The conclusion we have arrived at is, that the
colours of animals and plants are no mere accidents,
and are not created for our special benefit, but are
directly useful to their possessors, and have been
acquired because they are useful. Were any additional argument necessary, it would be easy to find
it in the fact that men are so far from being in
agreement as to what is and what is not beautiful,
that the ideal of one nation may be the horror of
another ; that a picture which an Art Committee
may select as beautiful, may appear to the public, for
whom it is purchased, as entirely destitute of beauty ;
that the various devices which savage races practise
in order to render themselves, as they consider,
beautiful, appear disfigurements to other nations.

So then it appears, on the one hand, that not only
is there no general agreement among mankind as to
what is beautiful, but that different nations, or the
same nation at different times, absolutely contradict
each other. Some other explanation is necessary of
the beautiful colours of animals and plants, and we
see what that explanation is in the great law of Utility,
expounded by the doctrine of Evolution, for the full
enunciation of which we are indebted to Darwin.

LECTURE VI

OBJECTIONS TO THE DARWINIAN THEORY

THE best possible mode of testing a theory is to consider the objections which have been raised against it. It is impossible for us to deal here with all the objections which have been put forward, but I propose to select those which appear the most important—*i.e.*, those which have been urged with the greatest force and persistency by men specially competent to deal with the subject. It is interesting to note that, in spite of the fierce storm of criticism to which the theory has been exposed, and the considerable amount of literature written on the subject—no small number of books having been written for the express purpose of "smashing Darwin"—yet nowhere are the objections and difficulties more clearly stated than by Darwin himself. Very few of any real importance have been added to the list given by Darwin, while he himself has indicated others that had escaped the notice of his opponents.

It seems strange to have to claim credit for candour; yet candour so striking as Darwin's does demand special and cordial recognition. Whether he was right or wrong in his conclusions,

Darwin simply sought to determine the truth, and was always ready to discuss and consider in detail even the most trivial and thoughtless objections.

MISSING LINKS.

The most popular objection, and in many ways the most famous, is that of the so-called Missing Links. If the present existing animals are descended from ancestors which were unlike them in former geological times, and if all animals are really akin or cousins, where are the intermediate forms, the missing links? These must on the theory of Evolution have existed. Can we produce them? or if not, can we give any reasonable explanation of our failure? We must at once admit that the demand is absolutely fair, and one which must be met. This question, although dealt with incidentally in former lectures, it is well to reconsider more directly. We have really two distinct problems to deal with, two kinds of links to be sought for—viz., (1) *Links between existing animals*, which must occur if the animals are akin to one another; (2) *Links between existing and extinct animals*. Let us consider these separately.

Links between the several kinds of existing animals. —Failure to find these has often resulted from misdirected efforts, from looking in the wrong direction. A straight line being the shortest distance between two points, it is commonly and not unnaturally assumed that the link must lie in the line connecting the two forms directly. True links are, however, not directly intermediate, the real relation being that of descent from a common ancestor, or branches of one stem.

Take for example the domestic pigeons. The blue rock is the common ancestor of all our domestic races, and is not in any sense intermediate between existing forms, such as a pouter and a fantail. Such intermediate forms have not existed at any time. So it is with ourselves: the real bond of union is through descent; two brothers are related, not directly, but through the parents; and with cousins the grandfather forms the real link.

These examples show that the actual links are not direct, but indirect; and that, unless careful, we may spend much time in looking for links where they could not exist, and overlook the real ones which are before our eyes all the time. This may be rendered deceptive by the occurrence of actual intermediate forms which are not true links, and against which we must be on our guard. An example will make this point clear.

The horse and the donkey are closely allied animals : midway between them is the mule, which shares the characters of both its parents. Yet this is clearly no true link, but an artificial unnatural creation that could not possibly have existed prior to either the horse or the donkey.

HORSE——MULE——DONKEY

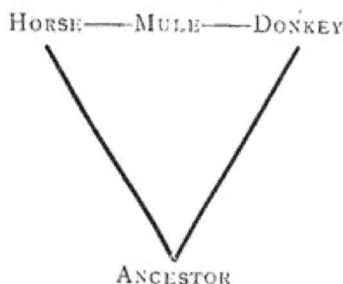

ANCESTOR

Developmental evidence we have found to be of
the utmost value, for the early stages in the development of animals are themselves the very links we
want ; sometimes distorted or modified, but usually
recognisable with sufficient care.   Take, for example, the fact of the prawn and the barnacle both
commencing life in the same form—*i.e.*, the Nauplius.
This is evidence of a most cogent character in
support of their descent from a common ancestor,
and here the Nauplius is the ancestral form, or link
from which both were derived.

Rudiments or vestiges also give most valuable
evidence.   For instance, the short tail of the crab ;
the splint bones of the horse's leg ; the mute letters
in our words : points which were fully considered in
a previous lecture.*

Links between the Present and the Past.

This was Cuvier's difficulty, and formed the basis
of an objection which was raised with fatal force.
If existing animals are descended from extinct forms
or fossils, why do the gaps appear so marked, and
where are the intermediate stages which should
exist ?

This question we have already dealt with at length
in a former lecture,† where we saw that the objection
could be met in its chief part by the *imperfection
of the geological record.*   We saw the extreme improbability that a continuous series of transitional forms
could be preserved, owing to the fact that only cer-

* See page 95.   † See page 60.

tain parts of animals can, in the ordinary course of
events, have any reasonable chance of being preserved
as fossils ; and furthermore, that only certain deposits, such as mud, are capable of preserving these
remains uninjured. " The crust of the earth," to
quote from Darwin, "with its embedded remains,
must not be looked at as a well-filled museum, but
as a poor collection made at hazard and at rare
intervals." We also, however, saw that in spite of
this imperfection of the record, several series of
fossil links have been obtained ; notably, in the
cases of the horse and *Paludina;* and that the
Archæopteryx is one of the most important links
known.

A further consideration in regard to fossil links
is that of the unlikeness between the dominant
forms of one age and those of succeeding ones : for
instance, between the reptiles of the secondary
period and the later forms. Undoubtedly, if we
regard these alone, there are great and apparently
abrupt gaps ; but we must not suppose fossil forms
all to stand in the direct line of ancestry of living
animals.

As it has been with animals, so it has been with men.
As we find in human history the Egyptian, Greek,
and Roman nations, each in turn dominant, so also
do we note that they are not lineal descendants one
of another, but collateral branches of the great
human family or tree, which in succession attained
maturity and gained dominion. Dominant races of
animals in successive geological ages have often died
out and left no descendants—great reptiles such as

the *Dinosaurus* for example. It will be these dominant races which will be the most likely to leave
fossil remains—as with the buildings and records
of man—and, especially if of large size, to give character to the age. Yet these do not usually stand in
the direct line of ancestry of living forms, which
are descended from collateral branches, which at
those times were insignificant. This consideration
explains at once the apparently sudden gaps, and
the difficulty in obtaining evidence of actual ancestors.
Dominant types often die out, and are succeeded by
others whose early history is difficult to unravel,
because overshadowed by the then dominant forms.

On the whole, therefore, the search for links is
not so hopeless as it appeared at first sight, when
we once realise what a link really is, and what we
have to look for. Additions are yearly, or almost
monthly, made to our knowledge of the former history of the earth, in the discovery of fossils, now
that their true importance is recognised.

Before concluding the subject of missing links
it will be well to refer briefly to that aspect of the
problem known familiarly as the "monkey question."
The Darwinian theory does most undoubtedly
imply that there is a blood-relationship between
man and the lower animals ; and it is also a most
undoubted fact that, of these animals, the anthropoid
apes—the orang, the chimpanzee, and the gorilla—
are those which are most closely allied to man.
Such being the case, much ingenuity has been
exercised in the search for "missing-links" which
will bridge over the gap between man and monkey,

and very bold statements have been based on the failure of these efforts.

Such ingenuity is, however, misdirected, and such efforts predestined to fail; for if what we have said above as to the real nature of links be true, it follows that the links between man and monkey will most certainly not be directly intermediate forms, any more than the link between the horse and donkey is a mule. From time to time we are provided with such alleged direct links, and it is a source of very legitimate amusement to a Darwinian to note the extreme anxiety people always show to prove that these waifs are really monkeys and not men, and the readiness with which they are able to accomplish this to their own satisfaction. And yet in all cases they have proved to be really human beings after all. Thus, of late years we have had a couple of idiot children spoken of as Aztecs, and a Siamese child called Krao, who differed from the rest of the family in nothing except unusual hairiness; not to mention other celebrities, all of whom were pronounced positively to be monkeys on their first appearance. Indeed, the only conclusion to be drawn from such cases, appears to be how very easy it is to persuade people to believe that human beings are monkeys.

Persistent Types.

Persistent Types, a group of cases which we have already considered,* have been brought forward

* See page 58.

as an objection to the Darwinian theory, on the
ground that, as the fittest survive, there should be a
continual improvement in the race; whereas, we
have seen forms such as *Lingula* and *Nautilus*
persisting apparently unchanged for periods of
enormous duration; or even through the whole time
of which we have any evidence in regard to life on
the earth. We must bear in mind, however, that
natural selection does not necessarily imply advance,
and is perfectly consistent with a stationary condition; for the fittest now may be the fittest many
years or ages afterwards. It is quite practicable for
particular forms to remain stationary for enormously
long periods, provided their environment, or at least
all features of the environment affecting them,
remain constant.

Persistent types are usually marine, for there the
conditions are more constant. They are very commonly found on sandy shores, in which they often
burrow; their tenacity of life and power of with-
standing injury are very great; they are usually also
unpalatable, and hence not eaten as food by other
animals. Their tenacity of life is well shown in the
case of *Lingula*, which was carried by Morse in his
pocket during a three days' journey across America
without coming to any harm; and *Balanoglossus* has
been known to live in a bucket of unaërated water
in a hot climate, with the hinder part of its body
completely macerated, and its branchial skeleton
exposed.

The persistence of lowly organised forms alongside more highly organised ones is often felt as a

difficulty ; for if the higher animals are descended from more lowly constituted ancestors, as we believe, and if these advances, these steps forward have been preserved because they were improvements, and because they gave advantage in the struggle for existence, how is it that we find the lowly organised forms still living alongside the higher and improved ones ?

The answer is that these lowly organised forms occupy places which cannot be filled by the higher forms ; as Wallace says : " There is no motive power to destroy or seriously modify them, and they have thus probably persisted, under slightly varying forms, through all. geologic time." Again, if we compare human history, we see that advance in civilisation does not involve the advance of all the members. For instance, some shops become larger and more pretentious, yet there are still plenty of places where small shops can survive, while there is no room for large ones. Again, some nations still exist which use bows and arrows, or slings and clubs, which are easily replaced and more suitable for their purposes than more modern implements of warfare.

DEGENERATION, OR RETROGRADE DEVELOPMENT.

It is very commonly assumed that, as in the struggle for existence, the fittest survive ; therefore each generation must be rather more highly or more perfectly organised, and fitter to survive than the preceding ; and that in all cases there must be a

steady and continuous, though slow, progress upwards. It is then asked, how is it that even at the
present day we find numerous representatives of the
simplest groups of animals living? And how is it
that we find many cases of degeneration—*i.e.*, of
animals which, in the early stage of their existence
—representing ancestral phases—are more highly
organised than in the adult condition?

Now, an animal may be placed under conditions
in which organs useful to its ancestors, and inherited
from them, may be no longer of service. Such
organs tend to become degenerate, persisting for a
time as vestigial structures, and ultimately perhaps
disappearing altogether. Of such cases of degeneration we meet with numerous examples, of which the
following are the most important:

(*a*) An animal fixed in the adult state, but free
when young: such as sponges, hydroids, corals,
polyzoa, oysters, and barnacles. This involves loss,
or modification, of the locomotive organs, and often
of the sense-organs as well.

(*b*) Parasites which live on or in other animals,
and of which *Sacculina* is a good example. In
these animals the whole body often becomes degenerate, the conditions of life rendering locomotor,
digestive, sensory, and other organs entirely useless.
In such cases, those forms which avoid the waste of
energy resulting from the formation and maintenance
of these organs will be most in harmony with their
surroundings. Parasitic worms, molluscs, &c., show
similar wholesale degeneration, and live immersed
in the body fluids of their victims.

The explanation of the extreme degeneration of parasites is that special food is required to meet the drain at the time of ripening of the eggs. For instance, in *Copepoda*, the female is alone parasitic, and that only at the time of laying eggs. The new phase intercalated in the life-history involves the necessity of laying more eggs, and there is greater difficulty in completing the ancestral history in individual development. This reacts on the parasitic stage, rendering it more important; more food is required, and hence further modification ensues.

(c) Special organs show signs of degeneration even in the highest animals, and give evidence of a former more perfect condition in their ancestral forms. This is seen in the eyes of the mole, and in many cave animals; in the splint bones of the horse, and in all the examples of rudiments or vestiges mentioned in a former lecture.*

In a sense, all the higher animals are degenerate; that is, they can be shown to possess certain organs in a less developed condition than their ancestors, or even in a rudimentary state. Thus, a crab, as compared with a lobster, is degenerate in regard to its tail; a horse, as compared with Hipparion, in regard to its outer toes. It is a mistake, however, to speak of a crab as a degenerate animal in comparison with a lobster, for an animal should only be spoken of as degenerate when the retrograde development has affected, not one or two organs only, but the totality of its organisation.

* See page 95.

No animal is at the top of the tree in all respects; man himself being primitive in retaining the full number of toes, and degenerate as regards his ear muscles. Care must also be taken not to speak of an animal as degenerate merely because it possesses organs less fully developed than allied animals. An organ is not degenerate unless its present possessor has it in a less perfect condition than its ancestors had. A man is not degenerate in the matter of the length of his neck as compared with a giraffe, nor as compared with an elephant in respect of the size of his front teeth, for neither elephant nor giraffe enters into the pedigree of man. A man is, however, degenerate, whoever his ancestors may have been, in regard to his ear muscles, for he possesses them in a rudimentary and functionless condition, which can only be explained by descent from some better equipped progenitor.

The theory of Natural Selection does not say that the ideally best survive, but those most in harmony with their surroundings for the time being. If these are of such a kind as to render certain organs useless, such as the eyes of cave dwellers, their possession is no longer an advantage, and the energy previously devoted to their production can be better utilised in other directions. Hence, though it is quite true that on the whole there has been a progress towards greater specialisation, and that differences between extreme groups are greater now than ever, yet there are many individual exceptions, and natural selection actually requires that there should be such exceptions.

THE ALLEGED USELESSNESS OF SMALL VARIATIONS.

This is an objection which has often been put forward. Admitting that no two animals are absolutely identical, it is urged that the differences are in most cases too small and too trivial to have the effect assigned to them; namely, to determine between survival or destruction. This is, however, a misapprehension, for if four out of five are to die, a very small matter may determine success or failure.

In a race for which there is but one prize, a victory by a short head is, so far as securing the prize is concerned, as conclusive as a win in a canter. Again, the whole theory and practice of the breeding of animals and plants afford absolute proof of the importance of attention to minute details, so slight as to escape the notice of all but the most skilful observers.

Think how in commerce a very small and subordinate point may determine survival; such as, for instance, the use of bye-products resulting from certain chemical manufactures, which had previously been neglected and regarded as waste products. Think what small events have decided the fate of battles and of nations. "The death of a man at a critical juncture, his disgust, his retreat, his disgrace, have brought innumerable calamities on a whole nation. A common soldier, a child, a girl at the door of an inn have changed the face of fortune and almost of nature."

The Difficulty with Regard to the Earliest Commencement of Organs.

This difficulty is a very serious one, for Natural Selection can only act on an organ after it has already attained sufficient size to be of practical importance and utility. Natural Selection accounts for any amount of modification in an organ when once established; modification in any direction, either of increase or decrease, but does not offer any explanation of the first appearance of such an organ. This is best understood by a few examples, showing the continuous preservation of a series of very minute variations.

1. *The wing of the Bat*, a flying mammal, is clearly a modified arm with great elongation of the fingers and webbing of the skin, which also extends from the side of the body and involves the hind-legs and tail (Fig. 28). It is easy to see that, when once established as a flying organ, Natural Selection would cause survival of those with the best wings, and so lead to gradual improvement and perfection of the wing. But how does the wing first commence?

Bats are a specialised group of mammals which must have been descended from non-flying ancestors. If the first commencement of the wing was a slight accidental elongation of the fingers, and a slight increase in the webbing, this would not give the power of flight, and would be of no use as a wing until it had attained a considerable size. In other words, *such an organ as a wing would in its earliest*

Pteropus (Fruit-eating Bat).

The central figure shows the skeleton with the wings (or *patagium*) outstretched.

The upper figure shows the favourite attitude of rest, with the wings folded up.

The lower figure shows the mode of progression along a branch, by means of the claws on the thumbs.

[N.B.—In this figure, the left-hand side represents the top, the right-hand side the bottom of the figure.]

stages be useless for the purpose which it ultimately fulfils.

This very important objection applies to a great number of cases, of which the origin of the wing is a typical one.

2. *The Origin of the Lung* is another example of the difficulty we are considering. The lung develops as a small saccular outgrowth from the throat, and it is quite unintelligible that a slight depression at the back of the mouth should have been preserved because it was useful for breathing air directly.

The explanations in these cases are good instances of many which are afforded by an important theory.

THE THEORY OF CHANGE OF FUNCTION.—This theory was suggested by Darwin, and afterwards developed more fully by Dohrn, as affording a possible solution of difficulties such as those we are considering. The principle is, that an organ may lose its original function, and yet persist because it is useful for another purpose—*i.e.*, that an organ may be used for two or more different purposes, one predominating at one time, another at another time; and, further, that structural modifications may ensue fitting the organ better for its adopted function.

This theory offers an explanation of the first commencement of the lung, which is shown to have arisen from the *swimming bladder* of fishes through change of function, and a series of forms is known to exist connecting the air bladder of fishes with the lung of the higher Vertebrates, which is undoubtedly the same organ.

The swimming bladder of most fish—the sturgeon for example—is a closed sac lying beneath the vertebral column, and is used for the purpose of flotation, to keep the back uppermost.   In many

FIG. 29.

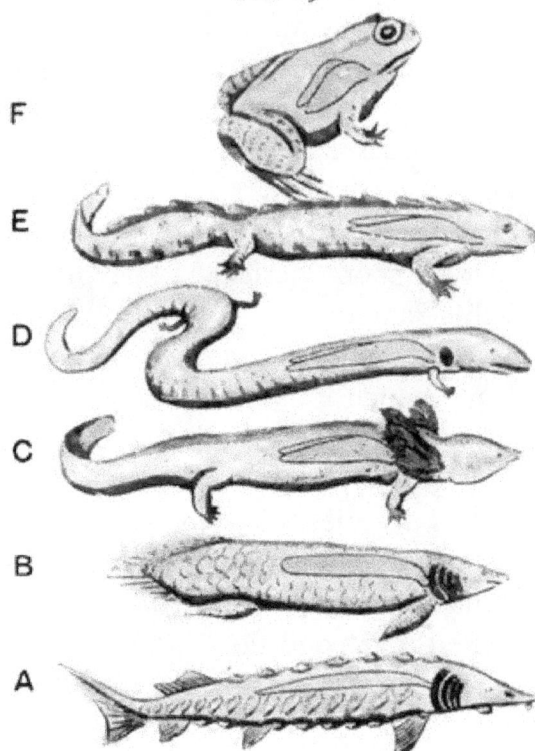

Diagram showing Evolution of Lung from Swim-bladder of Fish.

A, Sturgeon; B, *Ceratodus*; C, *Menobranchus*; D, *Amphiuma*; E, Newt; F, Frog.

fish it acquires a connection with some part of the alimentary canal, and then becomes an accessory breathing organ.

The mud-fish, *Ceratodus*, of Queensland and *Protopterus* of Africa, inhabit rivers which during

the dry seasons are apt to become dried up. These
animals lie buried in the mud for months, and can
live for a long time out of water, owing to the fact
that the swimming bladder is used as a lung, a
slight change in the circulation causing aërated blood
to be returned from it to the heart.

In *Menobranchus*, found in North America, both
lungs and gills are present throughout life, and it is
equally at home in water and on land. The lung
is better developed than in the Protopterus and
Ceratodus.

In *Amphiuma*, found in North American swamps,
the lungs are still more perfect. The gills are lost
but the gill-slits remain.

From this we reach the condition met with in the
newt and frog, which possess gills in the tadpole
stage, but lose them in the adult or lung-breathing
state. (See Fig. 29.)

We have thus a series of animals, all now living,
showing the actual transition from the swimming
bladder to the lung, and from the gill-breathing
to the lung-breathing condition ; and we further see
that the frog *actually repeats this history in its own
development.*

The explanation of the first commencement of the
bat's wing is more difficult, and there is still some
uncertainty about it. Let us consider another group
of Mammals, the SQUIRRELS, in which the finest gradation is known from animals with their tails slightly
flattened, the hind parts of their bodies wide, and the
skin of the flanks full, to the " flying squirrels," in
which the limbs and even the base of the tail are

united by a broad expanse of skin ; this fold of skin, acting like a parachute, enables them to glide through the air for a great distance from tree to tree, and so escape their enemies. Here each step is useful, and similar modifications are met with in other animals.

In *Galeopithecus*, the flying lemur of Borneo, the skin fold extends from the neck to the hand, thence to the foot and from this to the tail, and includes the limbs with the elongated fingers. From this Darwin suggested that the bat's wing could be derived by elongation of the fingers.

Further illustrations of the utility of imperfect wings to arboreal animals or fish are found. The flying frog of Borneo is a tree-frog with very long and fully webbed toes, which enable it to take long leaps in the air. The flying lizard (*Draco volans*) has the skin of the flanks supported by ribs. In both these cases there is no true power of flight, the action being that of a parachute. Of flying fish there are two chief groups, *Dactylopterus* (the gurnard), and *Exocoetus* (the flying herring). In both animals the pectoral fins are largely developed, and in the gurnard are almost certainly moved like wings.

There is no difficulty in understanding these cases as being acquired by Natural Selection, and the bat's wing may be not such a serious difficulty after all.

Other Examples of Change of Function.—A good instance of change of function is that of the *Hyomandibular* gill-cleft. The presence of gill-slits in

the early stages of development of the higher Vertebrates is very interesting, and the only possible
explanation of their presence is that their possessors
are descended from gill-breathing ancestors. These
gill-slits never bear gills, and all close up early with
the exception of one, the hyo-mandibular cleft, which
is preserved because its function is changed. This
cleft passes close to the ear, which is buried in the
side of the skull, and by remaining open becomes
advantageous for the purpose of hearing. So what
was once a gill-cleft has by change of function
become part of the organ of hearing.

The *electric organs* of some fish are good examples of change of function. These are always
formed by modification of muscular tissue, and in
all muscular contractions electric changes occur.
Perhaps this is an instance of a secondary function
becoming primary. In *Gymnotus*, the electric eel,
the electric organs lie just underneath the skin along
the sides of the tail. In *Malapterurus*, the electric
cat-fish, they extend over the whole body between
the skin and muscles, being especially developed at
the sides. In *Torpedo* they are used for stunning the
animals on which it preys, and also for purposes of
defence.

THE INSUFFICIENCY OF TIME.

Natural Selection is a slow process depending on
the gradual accumulation of small variations, for the
acquirement of which we have no actual standard of
time. Palæontology as yet tells us nothing as to the
origin of life, or even the origin of the large groups

of animals.   In the earliest fossil-bearing rocks we find the great groups typically represented, and in some cases, such as *Nautilus*, *Chiton*, and *Lingula*, by genera now living.   We seem driven to require an amount of time behind the Silurian period vastly greater than that which has elapsed since; how much no one can say.

On the other hand, physicists say we can only have a certain amount of time; for the earth is cooling, and it is a matter of calculation how long it has been cool enough for life to be possible. If evolution has really occurred, there must have been time, and the question for the biologist is whether there is evidence of evolution or not. Embryology suggests that the rate of change may be more rapid than is commonly suspected.

EVIDENCE OF DESIGN AND FORETHOUGHT.

This is a subject a little difficult to touch upon without trenching on matters which I wish to avoid. The evidence of adaptation of means to ends is especially manifest when we find contrivance or beauty. That there is harmony everywhere between animals and plants and their environment is undoubted; yet it appears to have escaped the notice of the objectors that this is the *very essence of the theory of Natural Selection*.   That there is evidence that any animals or plants are specially designed to satisfy the wants or to delight the senses of man is most absolutely denied; and could such cases be proved, they would be fatal to the whole theory.   In Nature those characters alone are preserved which are advantageous to the species.

INSTINCT.

Two objections have been raised with regard to instinct: (1) that it could not have been acquired through Natural Selection; (2) that it does not benefit its possessors, and therefore that its preservation is unintelligible.

Let us consider some examples of instinct. The eggs of butterflies and other insects are laid in places as safe as possible and near to their future food-supply. The butterfly never sees her young, and, feeding on the juices of flowers, can have no idea from her own experience as to the respective merits of different leaves ; yet she makes no mistake. Again, certain wasps sting the larvæ of beetles, so as to paralyse without killing them ; they then lay a single egg on the paralysed victim and leave it to its fate. The grub emerges and devours its prey, passing the winter in the pupa stage and emerging in the spring with the instincts of its parent. Here the individual wasp derives no advantage, but the gain to the species is enormous.

Preservation of habit, or instinct, is due to the fact that those individuals which take the greatest care to make provision for their young, will be most likely to give rise to offspring which will survive in the struggle for existence. Natural Selection will tend to preserve the instinct because it is *advantageous to the species, although of no benefit to the individual.*

LECTURE VII

In order to illustrate the conclusions arrived at in previous lectures, and to test the validity of the Darwinian theory, I propose to apply it, in a more detailed manner than we have yet considered, to one group of animals. The group selected is that of Vertebrates, as being a well-defined group, consisting of forms which are familiar and usually of comparatively large size ; and also for the reasons that the fossil forms in this group are numerous and characteristic, and the embryology of the group has been worked out with more detail than in the Invertebrates. Moreover, Vertebrates have a special interest from the fact that it is to this group that man himself belongs.

The problems we have before us are, first, to determine the mutual affinities of different groups now living ; secondly, to determine the relations of Vertebrates to other animals ; this being the less important of the two problems at the present time.

The evidence available is of three chief kinds :

(1) *Comparison of Structure.*—For example, a cat and a dog are clearly more closely allied to one another than is either of them to a fish or a bird.

(2) *Development.*—Here we obtain evidence from the Recapitulation theory, or the tendency of animals to repeat their past history in actual development.

(3) *Fossils.*—These afford the most valuable of all evidence, because it is the most direct and convincing, although at the same time the most fragmentary and incomplete.

Vertebrates form a good group for our purpose, inasmuch as in them evidence of all three kinds is available. The main characteristics of a typical Vertebrate are *the tubular nervous system*, forming the brain and spinal cord; the *notochord*, forming the main skeleton or backbone, and situated between the nervous system and the alimentary canal; the *myelonic eye*, or eye developed from the brain.

CLASSIFICATION OF VERTEBRATES.

I. CRANIOTA.—In these the skull and brain are present, and limbs nearly always so, or when absent have been clearly lost. The heart, liver, and other organs are well developed.

a. Pisces, or fish, are aquatic Vertebrates possessing gills, and provided with fins instead of limbs. They are rarely able to leave the water.

b. Amphibia, such as frogs, newts, and toads, are fresh-water or terrestrial. In early life they are aquatic, breathing by gills. Later on in life they may lose their gills and take to land life.

c. Reptilia. — These never have gills. They possess fore and hind limbs, furnished with fingers and toes, except in cases where they have been lost, as in the snakes. Among reptiles we find lizards, crocodiles, turtles, and snakes; and many extinct groups are known.

d. Aves, or birds, are characterised by the possession of wings and feathers, and have special modifications of the skeleton to aid in flight.

e. Mammalia.—These are typically terrestrial animals. Their main characteristics are the possession of hair, the presence of two pairs of limbs, furnished with claws or hoofs, and the fact that they do not lay eggs, but give birth to young which are suckled by milk glands.

II. ACRANIA.—This group contains a number of lowly organised forms, possessing no skull, no distinct brain, and no limbs. Their sense-organs and other parts, such as the heart and liver, are in a very primitive and simple condition. They are interesting as telling us in which direction to look for ancestors, for they show us that these may be destitute of limbs, eyes, ears, or backbone, or even without any skeleton. I propose to leave this group for a time, and first consider in more detail the several groups of *Craniota*.

PISCES.—The characteristic mode of breathing in fish is by gills. Gill-clefts have no meaning apart from their respiratory function, yet they are present in the early stages of development of all Vertebrates without exception. They are one of the most characteristic features of Vertebrates, and have a constant relation to the heart, blood-vessels, nerves, muscles, and skeleton. In fish and some Amphibia they are preserved in a functional state throughout life.

The inevitable conclusion is that, of the five groups of Craniate Vertebrates, fish are the most primitive; and that the other four groups are descended, if not from fish, at any rate from gill-breathing forms, aquatic and presumably fish-like. The evidence afforded by Palæontology does not help us much on this point, the oldest known Vertebrates being fishes from the lower Silurian deposits. Geology would favour an aquatic origin for Vertebrates as for other forms, the instability of the land in comparison with the sea being well known.

AMPHIBIA.—The chief point of interest in Amphibia is the well-known transitional series from the gill-breathing to the lung-breathing condition, which we have already dealt with at length when discussing the theory of change of function. Almost all Amphibia commence life as gill-breathers; but in one or two cases, such as *Hylodes*, where the eggs are not laid in water, owing to abundance of food-yolk the early stages are passed through before hatching, and, as in the higher Vertebrates, the gill-breathing stage is dropped out, though gill-clefts are developed (see Fig. 16).

The pedigree of Amphibia may be regarded as established, and is as follows : Amphibia are descended from fish, which migrated into rivers to avoid enemies and to obtain food. Owing to drought the air bladder became converted into a lung, and this at once conferred the power of going on land (see Fig. 29). This led to the conversion of a fin into a definite pentadactyle limb, a change the true nature of which is uncertain. The point of interest is, that every Amphibian repeats this history in its own development. Intermediate stages are not merely possible, but are known to actually occur at the present day, a most important point. The *Axolotl* is one of these, and is provided with large gills and also lungs ; the *Siren* is similar, and is found in the swamps of the Southern States of North America ; the mud-fish *Lepidosiren*, of Brazil, *Protopterus* of Africa, and *Ceratodus* of Queensland are other examples.

The origin of terrestrial Vertebrates is probably revealed to us in this way, and it must be considered that the fresh-water forms are descended from marine ancestors and the terrestrial forms from the fresh water.

REPTILIA.—Reptiles are an exceedingly abundant group, both absolutely and relatively, in the secondary period : a great number of families and orders are extinct. Although their actual origin is uncertain, their descent from gill-breathing ancestors is proved by their passing through the gill-cleft stage in their development. No reptiles, however, have gills.

It is uncertain whether reptiles are derived from Amphibia, or descended from gill-breathing ancestors directly. The gigantic *Ichthyosaurus* (Fig. 30) and *Plesiosaurus* (Fig. 31), which have more than five digits in the hand and foot, have been supposed to indicate an independent origin from fishes, but are more probably a reversion to the paddle-like form in an aquatic group. Moreover, the oldest reptiles yet obtained as fossils are pentadactyl, and not provided with paddles. Of

FIG. 30.

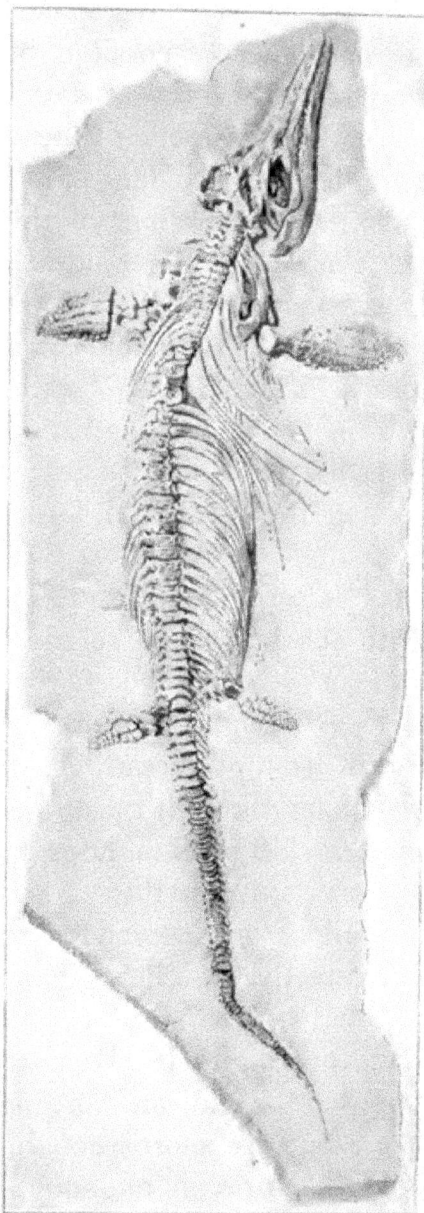

Ichthyosaurus.

FIG. 31.

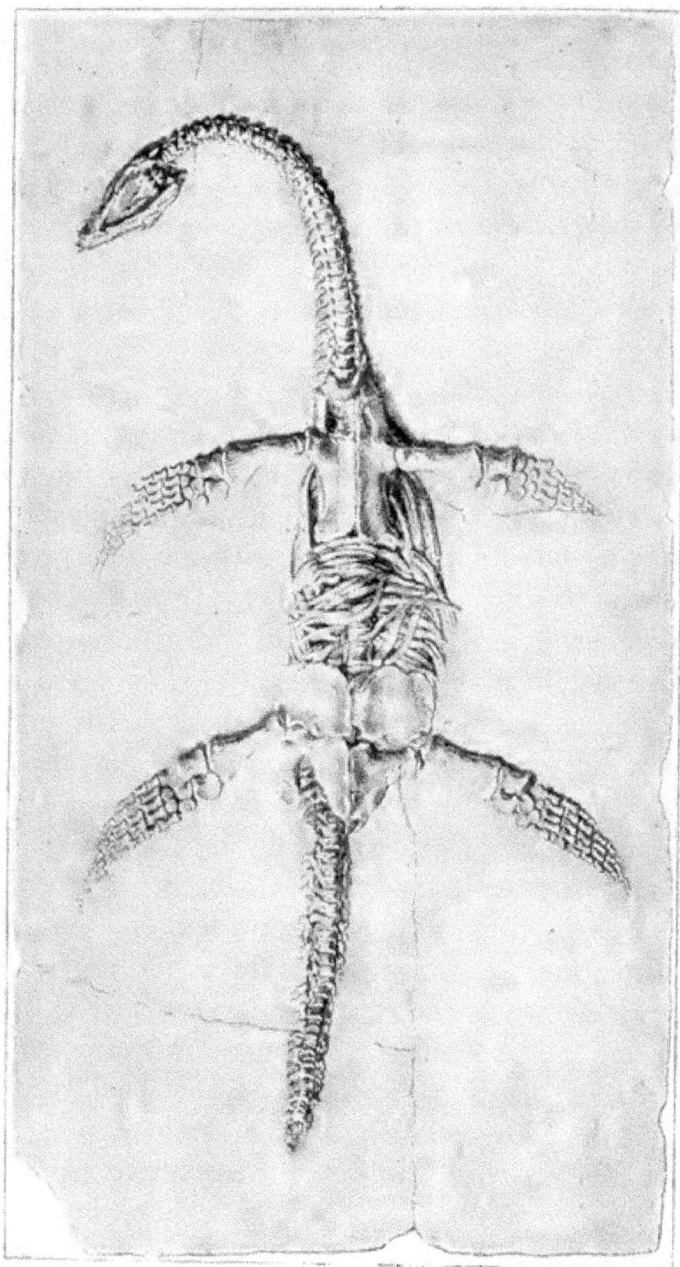

Plesiosaurus.

existing reptiles the lizards are the most primitive.
Snakes are comparatively recent tertiary descendants
of lizards, characterised by the absence or vestigial
condition of the limbs. Crocodiles and tortoises both
date back to a very remote period.

AVES.—Birds are a very special group, characterised by the possession of feathers, and the conversion
of the fore-limb into a wing. Their bones are very
light, with a marked tendency to fusion, as seen in
the metacarpal and metatarsal bones. The greater
number of their distinguishing points are obviously
correlated with the power of flight. Birds are warm-
blooded and really terrestrial, and form a compact,
well-developed group, at first sight very independent,
and having no obvious kinship with any one group
of Vertebrates rather than another. The affinities of
birds we have already discussed, and Palæontology
speaks very decidedly on the point. There are no
forms directly intermediate between birds and
reptiles, but such forms, we have seen, should not
exist.

We find reptiles which gradually lose the distinctive
reptilian characters, and birds which, as we proceed
backwards in time, gradually show less and less the
special features separating birds from reptiles.

This is shown in the case of the pelvic girdle; the
pubes in the crocodile pointing downwards and
forwards, in the bird downwards and backwards;
while in the *Dinosaurus* both processes are present,
and one is preserved in birds, the other in reptiles.
The *Dinosaurus* therefore forms a true link between
birds and reptiles; and this is supported by the

evidence given by development, indications of both processes being present in the development of the pelvic girdle in birds. Again, fossil birds, such as the *Archæopteryx*, are known which possess teeth and biconcave vertebræ, a long and many-jointed tail, and free metacarpal bones ; thus showing the convergence between birds and reptiles as we proceed backwards in time.

The actual origin of wings is still doubtful. The fore-limbs seem first to have become small, as is the case among the Dinosaurs, and then in some unknown way to have changed their function and become wings. Ostriches are said sometimes to give a possible clue, but more probably they are forms in which the power of flight has been lost, for wings are not essential to the bird type. The anatomy of birds supports the contention, for in most reptiles, although both left and right aortic arches persist, yet the right is much the larger. The presence of one condyle to the skull, the quadrate bone, and the absence of vertebral epiphyses are other points in favour of this view.

Evidence is also afforded by embryology. Birds pass through the gill-cleft stage, but have no gills ; therefore it is probably either to mammals or reptiles that we must look for allies. The large eggs of birds, the presence of egg-shells, and the details of their early development point to reptiles, as also does the development of the optic lobes of the brain, which though laterally placed in the adult bird, are dorsal in the embryo, and exactly like those of the lizard. The development of the metacarpal bones

and of the coccyx is also similar to that of reptiles;
and the development of feathers is comparable with
that of the scales of a reptile and very unlike the
hairs of a mammal.

MAMMALIA.—Mammals are characterised by being
warm-blooded animals with the following distinctive
points :—A heart with four cavities ; circular blood
corpuscles ; a diaphragm ; a left aortic arch ; a more
perfect brain, especially as regards commissures, than
the other groups ; no quadrate bone ; three special
ear bones ; two occipital condyles ; mammary glands
and hairs.   The anterior limbs are always present,
though the posterior may be absent.   There are
almost invariably seven cervical vertebræ.   Great
difference in size is met with among mammals, from
the harvest mouse to the whale.   They form by far
the most important group of all animals from an
economic standpoint, some of them serving as food,
some furnishing clothing, and others being used for
transport.

The geological history of mammals commences
with small forms found in the Trias, but comparatively scanty remains are found till the tertiary
period.   From tertiary times onwards, they are
found in three great areas, very distinct from one
another: (1) Australia ; (2) South America ; (3)
Europe, Asia, Africa, and North America.

The present distribution of Mammals is divided
into three great groups :

(1) *Monotremata :* a very small group of lowly-
organised animals, confined to the Australian region.
(2) *Marsupials :* formerly wide-spread, but now

confined to the Australian region and America ; the kangaroo and opossum are examples of this group. (3) *Placental mammals*, such as the horse, deer, lion, whale, bat, and monkey.

MONOTREMATA.—These are in many respects the most primitive, and it is therefore well to concentrate our attention on them.

FIG. 32.

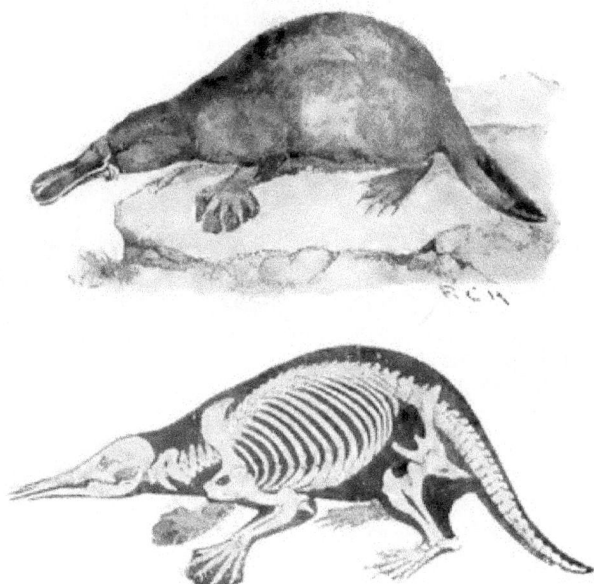

Ornithorhynchus.

The *Ornithorhynchus* (Fig. 32) is one of the most remarkable of animals, and was first described by Shaw in 1799.   The animal is from 18 to 20 inches in length, and is covered with short soft hairs.   It has a horny beak, something like that of a duck, and possesses teeth when young, although these are lost in the adult state.   It has small eyes, and no external ears.   The limbs are short and strong, and

in the fore foot the web extends considerably beyond
the fingers.   In the hind foot the nails are long and
curved, and in the male the heel has a spur an inch
long, at the apex of which opens a "poison"
gland.

The Ornithorhynchus is an aquatic animal living
in lakes and streams ; it swims and dives well, and
forms burrows, sometimes fifty feet long, on the banks
of the water in which it lives.   These burrows end

FIG. 33.

Echidna aculeata and *Proechidna Brujnii.*

in a dilated chamber which has two openings, one
below, the other above the water level ; in this
chamber the animal rolls itself up into a ball when
going to sleep.   It is found in Tasmania and the
southern and eastern parts of Australia.

Echidna, the spiny ant-eater (Fig. 33), is another
member of the group, and of this there are two
species found in Australia, Tasmania, and New
Guinea.   Its length is from a foot to 18 inches
or more.   It is covered with fur intermixed with

strong sharp-pointed spines like a hedgehog. It
has an elongated tapering snout, a long tongue used
for licking up ants, and no teeth. The feet are
provided with long and strong claws. Echidna
burrows very rapidly, and is mainly nocturnal in
habit. It can endure long fasts, and even exist for
a month or more without food.

Embryological Evidence.—Other Mammals bring
forth their young alive; but they are really developed
in the same manner as other Vertebrates, such as
the frog or bird, from a single cell or ovum; and
the embryo is retained within the body instead of
being laid.

The eggs of mammals are very small, that of the
rabbit being 0.116 mm. ($\frac{1}{220}$ inch) in diameter. The
curious point is, that the eggs of mammals develop
after the fashion of large eggs, and not in the
manner of small eggs. The difference in size of
eggs depends mainly on the amount of food yolk
contained in them; a small egg, such as that of the
frog, develops entirely and directly into the embryo;
a large egg, that of the chick for instance, is hin-
dered by the presence of food yolk, and becomes
constricted into two parts, the embryo and the yolk
sac. The egg of the rabbit develops as though it
had a large amount of food yolk, and forms a yolk-
sac. This fact is only intelligible on the view that
mammals are descended from ancestors having large-
yolked eggs. (Fig. 34.)

Monotremata are of special interest in this respect.
It was long believed that their eggs were laid as by
birds, and in 1829, Professor Grant described, on

the authority of a Mr. Holmes, the eggs of Ornithorhynchus as being ovoid in shape, equal at the two
ends, 1½ inches long and ¾ inch in diameter, with a
thin, whitish, transparent, calcareous shell. Of these
eggs, originally nine in number, four came to
England and two were deposited in the Manchester Museum, and labelled "eggs of duck-billed
Platypus."

In 1884 Monotremes were shown to be oviparous
by Mr. Caldwell, who was sent to Australia to study

FIG. 34.

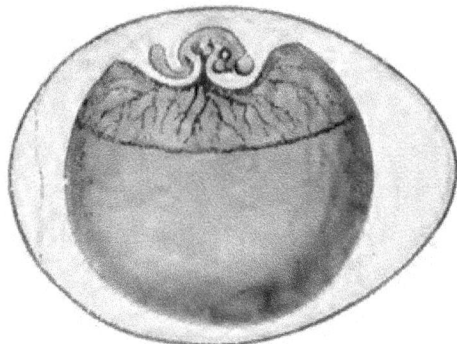

*Embryo Chick at the end of the fifth day of incubation, showing relation
of embryo to yolk-sac. Typical example of a large-yolked egg.*

the subject. The *Ornithorhynchus* lays two eggs at
a time, ¾ inch by ½ inch in size, enclosed in a strong
flexible white shell; *Echidna* lays only one egg.
The details of these eggs are not yet forthcoming,
but they strongly suggest affinities to reptiles.

Osteological Evidence.—The vertebræ have no
epiphyses; the sternal ribs are well ossified; the
mandible is devoid of an ascending ramus; the
humerus is very reptilian in character. There

are three distinct cartilage bones in the shoulder girdle—viz., coracoid, scapula, and pre-coracoid—agreeing with the Anomodont reptiles. Some of these extinct reptiles from the Permian (the top of the Palæozoic) and the Triassic deposits—*Anomodontia* or *Theromorpha*—in the character of the shoulder and hip girdles; in the perforation of the columella; in the relation and number of the bones of the hind foot, especially the astragalus and calcaneum; in the nature of the ribs and in other points, are in many ways intermediate between mammals and reptiles. They may possibly be the common ancestors of amphibia, reptiles, and mammals. This is a point of importance as showing a convergence towards a common ancestor.

We now enter on a second stage in our inquiry. We have found evidence in embryology of the descent of terrestrial Vertebrates from aquatic ancestors, in the presence of gill-slits, and in the actual transition shown us by Amphibians, such as the frog and newt. The geological evidence of the instability of land compared with the sea leads us to think of this as of general application, and not confined to Vertebrates, and that the origin of all the great groups of animals is to be looked for in the sea.

So far, we are led to regard fish as the most primitive of the five great groups of Vertebrates, and to view the remaining four as descended from fish or fish-like ancestors. Now, can we get any further? What is there behind fish, and can we trace the pedigree of Vertebrates further back?

This brings us to the consideration of the lower group of Vertebrates—ACRANIA.

Amphioxus (see Fig. 20).—This is a small fish-like animal from one and a half to two inches in length, living in the sand, and found at Messina, Naples, on our own shore, in Australia, and elsewhere. It possesses a notochord, a large number of gill-slits, and a central nervous system along its dorsal surface. Amphioxus is an animal of very primitive character, differing from other Vertebrates in the following points : The skeleton is of extreme simplicity, consisting of an elastic rod, the notochord. It has no limbs, no skull, no ribs, no brain, and no ears. The eye is a single pigmented spot at the anterior end of the nervous system. The alimentary canal is straight, and there are a hundred or more gill-slits (eight being the greatest number in fish). The heart is a straight tube, and the liver a simple diverticulum from the alimentary canal.

Yet Amphioxus is obviously a Vertebrate, for although it differs in almost all the above points from adult Vertebrates, it resembles the embryos of Vertebrates, and may be said to halt at an early stage in development. All Vertebrates pass through stages in which there are no limbs ; in which the notochord is the only skeletal structure, and the brain merely the anterior end of the neural tube ; in which paired eyes and ears are not yet formed ; in which the heart is a simple tube, the alimentary canal straight, and the liver a simple outgrowth from it.

The position of Amphioxus with regard to Verte-

brates is thus closely analogous to that of the
tadpole with regard to the frog ; a stage through
which the frog passes, but at which it does not halt.
The inference in the case of the frog is that frogs
are descended from tadpole-like—*i.e.*, fish-like—
ancestors.   Now, if this conclusion is rightly founded,
we may conclude that Vertebrates are descended
from ancestors like Amphioxus, and that of all living
animals Amphioxus most nearly represents the
common ancestor of Vertebrates.   The only alternative, and one that is urged by Dohrn, Lankester, and
others, is that Amphioxus is degenerate.   I think
this alternative wrong, and that there is no real
retrograde development, but that Amphioxus merely
stops at what is an early stage in the development
of the higher forms.   To my mind, Amphioxus is
one of the most important of Vertebrates, to be
regarded as shifting back the origin of Vertebrates
to an extraordinary extent, which is best realised by
saying that the differences between Amphioxus and
a fish are zoologically of far greater importance than
those between a fish and a mammal.

Ascidians, or Sea-squirts.   Some of these animals
are solitary and fixed, some in colonies, and others
free-swimming.   They have a covering or "test"
of cellulose with inhalent and exhalent apertures.
The pharynx has numerous gill-slits.   Water entering the inhalent aperture passes into the pharynx,
where the food-matter it contains is filtered from
it, and passing through the slits in the pharynx
escapes into the atrial cavity, and so out at the
exhalent aperture.   In Ascidians the nervous system

is represented merely by a single ganglion, and there
are no sense organs.   The notochord is absent.   In

FIG. 35.

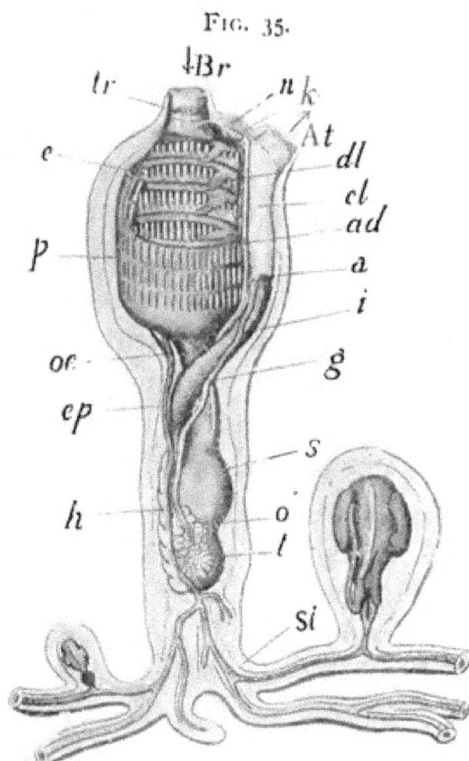

Clavelina, *a compound Ascidian, showing the anatomy of an adult member
of the colony, and two members in early stages of development (produced
asexually).   The members of the colony are connected together by the* stolon,
*which consists of an outer tube of ectoderm continuous with the outer layer
of the individuals; and an inner tube of endoderm, which is a direct pro-
longation of the so-called "epicardium."*

a, Anus ; *ad*, Dorsal nerve cord ; *at*, Atrial aperture ; *br*, Branchial aperture ; *cl*, Cloaca ; *dl*, Dorsal languets ; *e*, Endostyle ; *ep*, " Epicardium ";
g, Genital ducts ; *h*, Heart ; *i*, Intestine ; *k*, Sub-neural gland ; *n*, Nerve
ganglion ; *o*, Ovary ; *oe*, Œsophagus ; *p*, Pharynx, with numerous gill-
slits ; *s*, Stomach ; *st*, Stolon ; *t*, Testis.

(Clavelina is more primitive than most Ascidians in retaining the dorsal
nerve cord in the adult condition.)

fact, there is nothing in the adult Ascidian to indicate
Vertebrate affinities.   (Fig. 35.)

In their development, however, we find that they pass through the stage of a free-swimming larva, like a tadpole, which possesses a swimming tail, a nerve cord, a notochord, and an eye and ear. (See Fig. 8.)   This larva after a time becomes fixed ; the tail shrivels up and is absorbed ; the nervous system becomes reduced to a single ganglion ; the ear and eye become aborted, and the pharynx enlarges.   In fact, Ascidians are degenerate animals, which in their larval stage stand at about the same level as Amphioxus—if anything somewhat higher. Some Ascidians, such as *Appendicularia*, never get beyond the larval stage, and are found as minute free-swimming pelagic animals, with only one pair of gill clefts.

Concerning the zoological position of Ascidians, it is accepted by every one that they are Vertebrates, which is proved by their development.   Mistakes have, however, been made in assuming that they must be ancestral, and this mistake was made by Darwin himself. If ancestral, they should stop at what is a transitional stage in the development of higher animals, but they do not do this.   The resemblance is not between the adult Ascidian and the embryo Vertebrate, but between the embryo Ascidian and the embryo Vertebrate.   This in- dicates descent from some common ancestral stock, but not the descent of one group directly from the other.

Amphioxus and *Appendicularia* are the simplest existing forms with the characteristic vertebrate structure, and represent the nearest approach yet made towards the determination of the ancestry of Vertebrates. This suggests that the Vertebrate stem arose very early, and must not be derived from groups such as annelids, lobsters, spiders, and sea-urchins, each of which has been claimed as the ancestor of Vertebrates in recent years. Any attempt to get further back than this is mere speculation. The only available evidence is that of embryology, and that this is trustworthy we have abundant proof as regards the later stages of development. The Ascidian tadpole is shown by *Appendicularia* to be a possible adult, and indeed to correspond very closely to an existing adult animal. With regard to the earlier stages of development we have seen in a former lecture* that this will also apply, and that adult organisms are known corresponding to the several stages in the early development of Amphioxus which may possibly represent ancestral forms.

Thus, by the aid of anatomy, palæontology, and embryology we are able to define fairly clearly the broad lines of Vertebrate ancestry ; and the conclusion we arrive at is the usual one, viz., to emphasize the uselessness of search for directly intermediate links between existing forms, and to drive back the origin of this, the highest group of animals, to times earlier than those of the oldest fossiliferous rocks ; to show that our only hope of obtaining information concerning these first progenitors depends on the extent to which their existing descendants have preserved the record in

* See page 114.

their own development, and on our skill in deciphering this record.

THE DESCENT OF MAN.

And now we turn to the last stage in our inquiry, the zoological position of man.

Man is distinctly an animal—*i.e.*, neither a plant nor a mineral—requiring organised food, for which he is dependent on other animals or on plants. He is distinctly a vertebrate, as proved by his backbone, the relations of his nervous system, brain, heart, and sense-organs. Further, he clearly belongs to mammals—the presence of hairs instead of scales or feathers would alone be sufficient to show this; but he also possesses all the other characteristics of the group—viz., two condyles to the skull, seven cervical vertebræ, and a left aortic arch. Of the different groups of mammals, it is allowed on all hands that he is most closely allied to that of monkeys. The general shape of his body; the form of his limbs; the number and nature of his fingers and toes; the power of pronation and supination of the fore-arm; the shape of his head; the structure and size of his brain, and the form of his teeth, all prove this incontestably.

Further inquiry shows this correspondence to be a very close one. It is seen in every detail of structure of the human body, bone for bone, muscle for muscle, nerve for nerve, and even tooth for tooth. Man and monkey can be compared, and the most exact correspondence pointed out. This correspondence is so close that it is almost impracticable to find any constant points of difference of any value whatever. It has indeed been shown by Professor Huxley that the anatomical differences between man and the higher monkeys are markedly less than those between the higher and lower monkeys.

Again, as to development, man, like all other animals, commences as a single cell, and passes through a gill-cleft stage indistinguishable from that of other Vertebrates. He has, like other mammals, at one time two aortic arches, a right and a left, but during development he loses the right arch. His brain, eyes, and limbs are all formed in the same manner as in other mammals.

Rudimentary organs are also present in man, and are of the utmost value, because they are only explicable on the supposition that man is descended from some ancestor in which these organs were in a functional condition. For instance, the intrinsic muscles of the ear are present in man in an incompletely developed form, and in a condition in which they can be of no use to their possessors. So also with the *platysma* muscle, which, while extensively developed in some Vertebrates, such as horses, is in a comparatively rudimentary condition in man, and of scarcely any functional value. Again, the "wisdom teeth" form another example of rudimentary organs, since they are always cut long after the others, and sometimes never pierce the gums at all.

The tendency to *reversion* is also met with in man, and is seen in the more or less complete presence of two aortic arches in some cases, and in the occasional development of the muscles of the ear.

Like all higher animals man is primitive in some respects, for he is pentadactyl and plantigrade (*i.e.*, five-fingered and walks on the flat of the foot), and retains the power of supination and pronation of the fore-arm; these conditions being more primitive than those met with in other Vertebrates which are on the whole of lower organisation.

In fact, unless man wishes to continue going about the world stamped with living and palpable proof of his kinship with lower animals, he had better stop up his ears, or, still better, cut them off altogether; for so long as he bears at the side of his head those tell-tale flaps with their aborted and rudimentary muscles, so long as he hears by means of that slit, once a gill-cleft, now by change of function become an accessory organ of hearing, so long does he carry about in sight of all men sure proof of his relationship with lower, even with water-breathing animals.

Yet one can hardly recommend the operation, for if you were to remove one by one the various parts of your body which proclaim this kinship, you would get rid in succession of skin, muscles, nerves, bones, &c., and all that would be left in the end as man's special and peculiar possessions would be: (1) certain parts of his brain, and these only doubtfully; (2) the *extensor primi internodii pollicis* muscle, which straightens the first joint of the thumb; (3) the *peroneus tertius*, a small muscle in front of the lower part of the leg and ankle, inserted into the base of the little toe; (4) certain portions of other muscles.

Again, if we turn from bodily structure to the other characteristics of man, we find the same tendency to over-population, resulting in the same struggle for existence and the same survival of the fittest. Indeed, it was from the study of Malthus' " Essay on Population " that Darwin was led to the theory of Natural Selection. So it is with the history of the rise and fall of nations, with the evolution of human speech, customs, and clothing. All alike conform to the same laws as those regulating the structure and habits of other animals. And so with the influence of man on other animals ; the advent of man has simply been the arrival of another animal, better equipped and more cunning, more cruel than any other ; acting with supreme selfishness ; tolerating the existence of other animals only when they can be made subservient to his own wants or pleasures ; ruthlessly exterminating all that offend or thwart him. His very kindness is merely a nominal exception, for if perchance he appear kindly disposed to certain animals, it is only to satisfy his own selfish ends, that he may fleece them of their coats or pluck them of their feathers to adorn himself ; or to fatten them, that they may acquire a flavour more acceptable to his palate.

Application of the Darwinian Theory to the Language of Man.—Language has been said to be " the one great difference between man and brutes," and an " insurmountable obstacle to the theory of alliance by descent." This has been urged even by those who would accept the theory as applying to all other animals.

But has not language a history, has it not been evolved gradually, and is it not constantly, even daily, undergoing change? Is not this evolution, are not these changes of a nature precisely similar to those which have governed the animal kingdom in other branches, and have made it what it is at the present day? Did the English language suddenly appear, was it specially created, or was it gradually evolved by slow modification of other tongues, such as Latin and Saxon? Why do modern languages— Italian, French, Spanish and Portuguese—have so many words closely similar or identical? Is it not owing to descent from a common Latin ancestor?

Again, Latin and Greek have many likenesses to one another, and with Celtic, Teutonic, Slavonic, and the ancient dialects of India and Persia, may be all regarded as descended from one common Indo-European or Aryan stock. So Hebrew, Arabic, and Syriac form a Semitic group. Moreover, "suppose," says Max Müller, "we had no remnant of Latin; suppose the very existence of Rome and of Latin were unknown to us, we might still prove on the evidence of the six Romance dialects that there must have been a time when these dialects formed the language of a small settlement; nay, by collecting the words which all these dialects share in common, we might to a certain extent reconstruct the original language, and draw a sketch of the state of civilisation as reflected by these common words."

Moreover, evidence of recapitulation is shown by the way in which a child learns to speak its own language, and "we know for certain that an English child, if left to itself, would never begin to speak English." We also see examples of rudimentary organs well illustrated by the silent letters in words such as doubt, reign, feign, debt, and answer. Persistent types in language are also met with. "The language which the Norwegian refugees brought to Iceland has remained almost the same for seven centuries; whereas, on its native soil and surrounded by local dialects it has grown into two distinct languages—Swedish and Danish."

Constant change is found in words at different periods. We can read Milton, Bacon, Shakespeare, and Hooker, though conscious of unfamiliar words and obsolete expressions; we can make out Wycliffe and Chaucer; but when we come to the English of the thirteenth century we can but guess its meaning. A Bible glossary shows that since the year 1611, three hundred and eighty-eight words, or one-fifteenth of the whole number used, have become obsolete : and, on the other hand, new words are constantly being added.

Conclusion.

This is the precise position which I have endeavoured to establish : that there are causes which will account for what we find—for the structure, language, and habits of man ; causes which have been in existence ever since life began, and causes which must have tended in this direction.

Whether there is anything further than this, whether man has other attributes, either peculiar to himself or held by him in common with other animals ; whether there are attributes that cannot be explained by these laws, is a question with which we have no concern here.   Science has nothing to do with such matters, and has nothing to say either for or against them.

Such is the doctrine of Evolution as applied to man, and I would ask you, Is there anything humiliating in this ?   Surely it cannot be more degrading to have risen than to have fallen.   Surely the true interest of life lies in the future rather than the past ; in the possibility of further achievements ; in there being work ahead for us to do.   It is in the consciousness that we now possess the key that will compel the past to yield up its secrets, and that opens to us unbounded possibilities in the future ; it is in the conviction that there is a reason in and for everything, and that it is within our powers to determine that reason, that we find the great charm and attractiveness of the Darwinian Theory.

LECTURE VIII

THE LIFE AND WORK OF DARWIN

HITHERTO we have been concerned with the great theory with which Darwin's name is inseparably connected ; we have dealt successively with its birth and maturation ; we have tried to form some idea of its wide-reaching influence, and of the effect which it has had, not merely on biological thought, but on other fields of science, literature, and art, and branches of knowledge apparently widely remote. We have seen how this theory has knit together human knowledge, giving the word *history* a new, a wider, a more wonderful significance than was possible before. We have dealt, I admit too briefly, with the main objections to the theory, and have taken a single instance in detail as a test and as an example of methods.

There is no more fitting way of concluding this series of lectures than by giving an outline of the life and work of the man to whom this great advance, this opening up of new fields, this widening of human interests and human powers, is due. Concerning his life, the progress of the events through which such results were obtained, the methods by which success was compelled, the successive steps in the development and ripening of the great theory; all these must have much of interest, much that will repay the hearing. Concerning his works, although the "Origin of Species" remains by far his greatest achievement, yet it must not be supposed that Darwin gained one great victory and then rested. No more conscientiously industrious man ever lived; and besides his masterpiece he has left us a great series of books, each dealing with a separate group of problems in animal or in plant life; each based on long-continued and scrupulously exact observations; each breaking entirely new ground; and each contributing powerfully to the advancement and widening of knowledge. While never forgetting that the "Origin of Species" stands foremost, it is well that the other works should not be overlooked; for had the "Origin of Species" never been written, these works—as yet hardly mentioned in our course —would have given Darwin a foremost place among the biologists of all nations and of all ages.

FAMILY HISTORY.

Charles Darwin was born on February 12th, 1809, at Shrewsbury. His mother was a daughter of Josiah Wedgwood, the founder of the great pottery works at Etruria. His father, Robert Waring Darwin, was a physician in large practice at Shrewsbury; a man of marked individuality of character, a quick and acute observer, with a great power of reading character and of winning the confidence of his patients. He was highly esteemed for his skill in diagnosis, but was not a man of special scientific ability. By his large practice he accumulated a considerable fortune, and was able to leave his children in easy circumstances. Darwin's grand-father, Erasmus Darwin, 1731–1802, was a physician of great repute, and the author of "Zoonomia," an ambitious treatise, showing extensive rather than profound acquaintance with natural phenomena; containing many bold, ingenious, and at times fantastic speculations. He was also the author of numerous and voluminous poetical works. He pro-pounded a hypothesis as to the manner in which species of animals and plants have acquired their character, which is identical in principle with that subsequently rendered famous by Lamarck.

Charles Darwin in his childhood and youth gave no indication that he would do anything out of the common. He was a strong, well-grown, active lad, interested keenly in field sports. "In fact," says Huxley, "the prognostications of the educational authorities into whose hands he first fell were distinctly unfavourable, and they counted the only boy of ori-ginal genius who is known to have come under their hands as no better than a dunce." From 1818 to 1825 Darwin was at Shrewsbury School under Dr. Butler, leaving at the age of sixteen. "Nothing could have been worse," he says, "for the development of my mind than Dr. Butler's school, as it was strictly classical, nothing else being taught except a little ancient geography and history. The school as a means of education to me was simply a blank." Yet, not incapable of appreciation, he writes : "The

sole pleasure I ever received from such studies was
from some of the Odes of Horace, which I admired
greatly." He also says : "I used to sit for hours
reading the historical plays of Shakespeare."

He was interested in chemistry, and fond of
making experiments with his brother in the tool-
house at home. He writes : " The subject interested
me greatly, and we often used to go on working till
rather late at night." This became known at the
school, and earned for him from his schoolfellows
the nickname of " Gas " ; and from the head-master
a public rebuke for "wasting his time on such
useless subjects."

Doing no good at school, he was sent to Edinburgh in 1825, with the intention of studying
medicine. This, however, was not much of an
improvement, for, as he writes : " The instruction
at Edinburgh was altogether by lectures, and these
were intolerably dull." The Professor of Anatomy
made his lectures "as dull as he was himself"; and the
lectures on Materia Medica were "something fearful
to remember," even forty years later. But the
climax seems to have been attained by the Professor
of Geology and Zoology, whose prælections were so
"incredibly dull" that they produced in their hearer
the determination—fortunately for the world not
adhered to—"never to read a book on geology, or
in any way to study the science." He, however,
became acquainted with some good practical naturalists, and got lessons in bird-stuffing, and also became
a good shot.

After two sessions at Edinburgh his father

decided that he had little or no taste for the life of a physician, and fearing that he might sink into an idle sporting man, proposed that he should go to an English University with the view of becoming a clergyman. So far as the direct results of academic training were concerned, the change of Universities was hardly a success, for he writes: "During the three years which I spent at Cambridge my time was wasted, so far as the academical studies were concerned, as completely as at Edinburgh and at school."

And yet it would appear that the fault lay rather with the method than the man; for he speaks of Algebra and Euclid as giving him much pleasure. He also studied Paley's "Evidences" very thoroughly, and expresses himself as being much delighted with the logic of the book, and charmed by the long line of argumentation. He was fond of outdoor sports, especially riding and shooting.

He was devoted to collecting insects, or, as he expresses it, "mad on beetles." This was a point of much importance, as it brought him in contact with Henslow, the Professor of Botany, a man of singularly extensive acquirements, who took a keen pleasure in gathering young men around him, and in acting as their counsellor and friend: "A man of winning and courteous manners; free from every tinge of vanity or other petty feeling."

At Henslow's advice Darwin was led to break his vow never to touch geology, and through him he obtained permission to accompany Professor Sedgwick on a geological excursion in Wales, by which

he gained the practical knowledge which he was so
soon to put to the test. It was Henslow who
advised him to read Lyell's " Principles of Geology,"
which had just been published in 1830, advising him,
however, on no account to adopt Lyell's general
views ; a piece of advice which he promptly
neglected, for, as we have seen, it was by the
unflinching application of Lyell's ideas and methods
to Biology that Darwin was led to his greatest
results. Finally, it was Henslow who obtained for
him the permission to accompany the *Beagle* on
her memorable voyage. This, he says, was "the
turning-point of my life."

The *Voyage of the Beagle* occupied from 1831
to 1836, Darwin being then twenty-two years
of age. In the autumn of 1831 it was decided
by the Government to send a ten-gun brig of
242 tons burden, under Captain Fitzroy, to
complete the unfinished survey of Patagonia and
Tierra del Fuego, to map out the shores of Chili
and Peru, to visit several of the Pacific archi-
pelagoes, and to carry a chain of chronometrical
measurements round the world. This was essentially a scientific expedition, the captain, afterwards
Admiral Fitzroy, being himself an accomplished and
highly-trained officer, and famous as a meteorologist.
Anxious to be accompanied by a competent naturalist,
to collect animals and plants, he generously offered
to give up part of his own cabin accommodation.
On Henslow's recommendation this was offered to
Darwin, who was eager to accept it. His father,
however, objected strongly, adding : "If you can

find any man of common sense who advises you to
go. I will give my consent." His uncle, Josiah
Wedgwood, strongly urged him to accept, where-
upon his father gave his consent. He, however,
narrowly escaped rejection, Fitzroy doubting
whether a man with such a shaped nose could
possess sufficient energy and determination for the
voyage !

This voyage, originally intended to last two years,
and ultimately extended to five years, started from
Devonport on the 27th of December 1831, returning
to Falmouth on October 2nd, 1836. Darwin writes :
"This was by far the most important event in my
life, and has determined my whole career." It was
during this time that he acquired habits of energetic
industry and concentrated attention. He collected
largely, and from his collections laid the foundation
of his great work.

It is necessary to bear in mind that this was
essentially a surveying voyage, and the bulk of the
time was occupied in a detailed survey of the east,
south, and to a less extent the west coast of South
America ; involving slow work and repetition of
much of it, the ground having to be covered more
than once in difficult places.

The *Beagle* left Devonport on December 27th,
1831, calling at the Cape Verde Islands and
St. Paul's Rocks, and reached Bahia on February
29th, 1832. After a short stay she proceeded southwards to Rio and Monte Video. The next three
years were spent in the special work of surveying :
nearly two years on the east coast off Tierra del

Fuego and the Falkland Islands, and rather more than a year on the western coast. Darwin went partly with the ship, but spent long periods on shore, travelling over much country, collecting, observing, and thinking.

Darwin's most important overland journey was in 1833 from Rio Negro to Bahia Blanca, and thence five hundred miles further on to Buenos Ayres. During this journey over Pampas he discovered the remains of vast numbers of extinct animals, some of enormous dimensions, such as the *Megatherium* and the *Mylodon*, which were found in gravel fifteen to twenty feet above the sea level near Bahia Blanca. It was here that Darwin was much struck with the relations between living and extinct forms. "This wonderful relationship," he writes, "in the same continent between the dead and the living, will, I do not doubt, hereafter throw more light on the appearance of organic beings on our earth, and their disappearance from it, than any other class of facts." He noticed the replacement of huge extinct forms by unlike yet allied forms, and that species are replaced, but by allied species. Darwin was impressed rather with their resemblances than their differences, whereas Cuvier, we saw, was most struck with the differences. He was also much impressed with the evidence of changes in the land in recent times, a point which laid the foundation of his theory of coral reefs. At Tierra del Fuego he was struck with the characters of savage races, and noted the almost entire absence of everything which we regard as characteristically human.

The Galapagos Islands, five or six hundred miles west of America, on the equator, are all volcanic, and therefore presumably recent in a geological sense. Nearly all the animals found here are peculiar to the islands, and different islands have their own fauna; yet these are more nearly akin to those of South America than to any other forms. In 1835, Darwin visited these islands, and set himself to discover the reason of this. Writing in 1837, after his return, he notes : " In July opened first note-book on Transmutation of Species. Had been greatly struck from about the month of previous March on character of South American fossils, and species on Galapagos archipelago. These facts (especially latter) origin of all my views."

From 1836 to 1842, the first six years after his return from the *Beagle* voyage, much time was spent at first in unpacking specimens and distributing the collections among specialists. The geological specimens he reserved for his own share. The publication of detailed results occupied much time, and in 1839 the " Naturalists' Voyage " was published, the second edition appearing in 1845 as the " Narration of the Voyage of the *Beagle*." The third edition in its present form appeared in 1860.

The geology of the voyage was written by Darwin himself, and consisted of three parts, the best known of which dealt with the *structure and distribution of coral reefs*. " No other work of mine," he says, " was begun in so deductive a spirit as this, for the whole theory was thought out on the west coast of South America, before I had seen a true coral reef."

The years between 1836 and 1842 were marked by the gradual appearance of that weakness of health which ultimately forced him to leave London, and take up his abode for the rest of his life in a quiet country-house. This greatly discouraged him, and threatened thus early to become permanent. In 1839 he married his cousin Emma Wedgwood, and lived for three years in Upper Gower Street, then suffering much from illness. In 1842 he purchased a house at Down, near Beckenham, where the rest of his life was spent, and the greater part of his work accomplished.

The "Origin of Species" was his first and greatest work, the "chief work of my life." We have already sketched the history of the development of the theory, and may now consider it from the more personal point of view. During the voyage of the *Beagle*, Darwin was led to think much on the subject ; and on his return, as soon as he had leisure, returned to it. During the voyage he "believed in the permanence of species," though experiencing occasionally "vague doubts." On his return in 1836, while preparing his journal, he noted how many facts indicated the common descent of species.

On the 1st of July 1837 he opened his first note-book to record any facts bearing on the transmutation of species, and "did not become convinced that species were mutable until I think two or three years had elapsed."

Darwin was most impressed by the South American fossils, and by geographical distribution, especially in the Galapagos Islands. The problems which occurred to him were: "Why are the animals of the latest geological epoch in South America similar in *facies* to those which exist in the same region at the present day, and yet specifically and generically distinct?" And, "Why are the animals and plants of the Galapagos Archipelago so like those of South America, and yet different from them? Why are those of the several islets more or less different from one another?"

These problems were only explicable on the assumption of modification, and in order to explain the cause of these modifications he turned to the only certainly known examples of descent with modification—viz., those presented by domestic animals and cultivated plants. The details of these he worked up and experimented upon in a much more thorough manner than his predecessors, especially in regard to pigeons. He soon perceived "that selection was the keystone of the main success in making useful races of animals and plants"; but says: "how selection could be applied to organisms living in a state of nature remained for some time a mystery to me."

"In October 1838," he writes, "that is, fifteen months after I had begun my systematic inquiry, I happened to read for amusement, 'Malthus on Population,' and being well prepared to appreciate the struggle for existence which everywhere goes on from long-continued observations of the habits of animals and plants, it at once struck me that under these circumstances favourable variations would tend to be preserved, and unfavourable ones to be destroyed. *The result of this would be the formation of new species.* Here, then, I had at last got a theory by which to work; but I was so anxious to avoid prejudice that I determined not for some time to write even the briefest sketch of it. In June 1842, I first allowed myself the satisfaction of writing a very brief abstract of my theory in pencil, in thirty-five pages, and this was enlarged during the summer of 1844 into one of 230 pages."

Not till 1858 was the theory published, and then only on pressure of the strongest character being brought to bear on him. On June 18th, 1858, Darwin, having convinced himself and accumulated the evidence and proofs he wanted, was at last at work on the book, when he received, most unexpectedly, Wallace's MS. from Ternate, in which the Theory of Natural Selection was set forth clearly and decisively, almost in Darwin's own words. Darwin wished to publish Wallace's paper without reference to his own work; but at the urgent solicitation of Lyell and Hooker he consented to allow extracts from his own MS. of 1844, together with a letter to Asa Gray of 1857, to be read before the Linnean Society on July 1st, 1858.

Darwin was at this time forty-nine years of age, and Wallace thirty-five. On Darwin's part this publication was the result of twenty-one years of deliberate work; and of views formed nineteen years beforehand, and actually written out in MS. of 230 pages fourteen years previously. The dual authorship of the "Theory," and its simultaneous announcement from opposite sides of the world, were causes for sincere congratulation. It was fortunate for Darwin, in causing him to publish his views more speedily, and in a more condensed and attractive form than he originally purposed, and as leaving him at liberty for further work. It was also fortunate for Wallace in securing cordial and sympathetic recognition in the most gratifying manner of his independent discovery. Finally, it was fortunate for the world, and a lesson for all time to come, of how an emergency, involving the tenderest susceptibilities of scientific reputation, can be treated so as to redound to the infinite and lasting credit of all concerned.

With regard to Wallace, it is interesting to know the immediate causes which suggested the theory, especially in view of Darwin's history, and I cannot do better than quote Wallace's own words :

"In February 1858 I was suffering from a rather severe attack of intermittent fever at Ternate, in the Moluccas ; and one day, while lying on my bed during the cold fit, wrapped in blankets, though the thermometer was at 88° Fahr., the problem again presented itself to me, and something led me to think of the 'positive checks' described by Malthus in his 'Essay on Population,' a work I had read several years before, and which had made a deep and permanent impression on my mind. These checks—war, disease, famine, and the like— must, it occurred to me, act on animals as well as man. Then I thought of the enormously rapid multiplication of animals, causing these checks to be much more effective in them than in the case of man ; and while pondering vaguely on this fact there suddenly flashed upon me the *idea* of the survival of the fittest—that the individuals removed by these checks must be on the whole inferior to those that survived. In the two hours that elapsed before my ague fit was over, I had thought out almost the whole of the theory ; and the same evening I sketched the draft of my paper, and in the two succeeding evenings wrote it out in full, and sent it by the next post to Mr. Darwin."

Thus we see that Malthus was the turning-point with both authors ; and all historically inclined will place this famous essay on their shelves, alongside the works of Darwin and Wallace. To me it has always been a striking fact, that with both authors, and perfectly independently, the turning-point in the argument should have been the application to the whole animal world of principles already established and accepted in regard to man. This is curiously significant in view of the objection, so often and so ignorantly brought against Darwinism, that it is degrading, because it applies to man laws governing the structure and habits of animals.

From 1846 to 1854 Darwin devoted himself to a monograph on Barnacles, consisting of four large volumes, two on recent and two on fossil species : a heavy task which he was led to under-take, largely, through "a sense of presumption in accumulating facts and speculating on the subject of variation without having worked out my due share of species." "No one," he says, "has a right to examine the question of species who has not minutely described many."

Subsequent to the publication of the "Origin of Species" much time was taken up by successive editions of the work, and much lost through ill-health. Darwin himself described his books as the "milestones of my life"; and they fall into two great groups—those completing the "Origin of Species," and those on more or less independent lines, such as the great Botanical Series.

I. WORKS COMPLETING THE "ORIGIN OF SPECIES."

This work was itself an "abstract," and this was the title he proposed to give it, having originally designed it to be of much greater length. However, yielding to the publishers, he produced it in its present form.

In 1868 appeared "Variations of Animals and Plants under Domestication," in which a detailed account was given of artificial breeding; and this was intended to be followed by two other works dealing with variation, heredity, embryology, geographical distribution, &c., in similar detail; but these were never written.

The detailed application of the theory to man was inevitable; and in the first edition of the "Origin of Species" he says: "In the distant future I see open fields for far more important researches. Psychology will be based on a new foundation, that of the necessary acquirement of each mental power and capacity by graduation. Light will be thrown on the

origin of man and his history." In 1871 the
"Descent of Man" was published, and in 1872
"The Expression of the Emotions," which was
originally intended to be only a chapter in the
"Descent of Man."

II. THE SERIES OF BOTANICAL WORKS.

These dealt with the development of special
questions and problems arising in direct connection
with the "Origin of Species."

In 1862 appeared the "Various Contrivances by
which Orchids are Fertilised by Insects," of which
Professor Huxley says: "Whether we regard its
theoretical significance, the excellence of the observations, and the ingenuity of the reasonings which
it records, or the prodigious mass of subsequent
investigation of which it has been the parent, it has
no superior in point of importance."

From the first, Darwin was convinced that no
theory could be satisfactory which did not explain
the way in which mechanisms, involving adaptation
of structure and function to the performance of
certain operations, had come about. In 1793
Sprengel had established the fact that in a large
number of cases a flower is a piece of mechanism,
the object of which is to convert insect visitors into
agents of fertilisation. What Sprengel did not do
was to show that plants provided with such flowers
gained any advantage thereby. Darwin worked at
this subject for many years, from 1839 onwards, and
showed cross-fertilisation to be favourable, and in

many cases essential, to the fertility of the plant and the vigour of the offspring ; and that all mechanisms favouring cross, and hindering self-fertilisation, give an advantage, and are hence preserved and improved through Natural Selection.

Orchids form an excellent group for the study of these mechanisms of cross-fertilisation, extraordinary modifications of which are found. The flowers are large and conspicuous, and many of them of singular form. The method of cross-fertilisation in *Orchis mascula* we considered in detail in a former lecture.* We must remember that all these details were worked out by Darwin before he had himself seen any insects visit the particular orchids which he described, the necessary confirmation being supplied in 1873 by Hermann Müller in his work on the " Fertilisation of Flowers."

In fine, an extraordinary diversity of devices, of marvellous interest, for ensuring cross-fertilisation, is met with among orchids. In some the pollen cases explode on being touched, and shoot their pollen at the insect. Some flowers are adapted for particular insects, and failure to become acclimatised is now recognised in many cases as being due, not to the plants being unable to live, but to the absence of the proper insects which alone can effect fertilisation, these as a rule being bees and butterflies. This subject may well be described as the Romance of Natural History, and Darwin says : " I never was more interested in any subject in all my life than in this of orchids."

* See p. 146.

In 1876 appeared "Cross and Self-Fertilisation in Plants," the outcome of a great series of laborious and difficult experiments on the fertilisation of plants, which occupied him for eleven years. The book on orchids showed how perfect are the means for ensuring cross-fertilisation ; the new book demonstrated how important are the results. The investigations of which this book was the outcome were commenced with a simple experiment made for quite another purpose. Darwin raised two large beds, close together, of cross-fertilised and self-fertilised seedlings of *Linaria vulgaris*, the common toad-flax, and his attention was aroused by the fact that the cross-fertilised plants, when fully grown, were plainly taller and more vigorous than the self-fertilised ones.

In 1877 the "Forms of Flowers" was written, being the development of a short paper read before the Linnean Society in 1862 on the two forms, or dimorphic conditions, of *Primula :* one of which has short stamens situated in the middle of the tube of the corolla, and a long style, the stigma of which is on a level with the open flower. The other form has long stamens reaching to the centre of the flower, while the style is short and the stigma half-way down the corolla, at the same level as the stamens of the other form. (See Fig. 36.) Darwin showed that these flowers are barren if insects are prevented from visiting them, and further, that each form is almost sterile when fertilised by its own pollen, but each is fertile when fertilised with the pollen of the other. By referring to the figures it will be seen

that insects visiting the flowers will carry pollen from
the long anthers of the short-styled form to the
stigma of the long-styled form, but would not reach
the stigma of the short-styled form.   Darwin showed
that in this way more seeds were produced than by
any other of the four possible unions.   Thus,

FIG. 36.

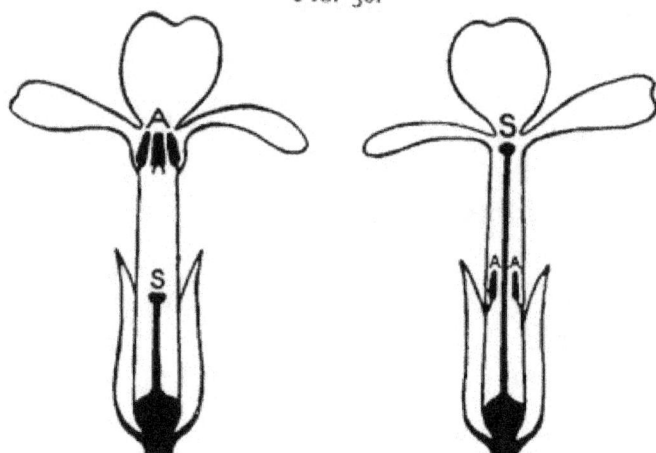

Short-styled flower.   Long-styled flower.

Diagram of dimorphic flowers of *Primula*.

a, Anthers.   s, Stigma.

although both forms are hermaphrodite, they bear
the same relation to each other as do the two sexes
of an ordinary animal.

A few more botanical works must be mentioned,
published in 1875, each of which was the outcome
of many years of laborious observations and experiments; each breaking new ground and giving
results of great importance and interest.

In "Climbing Plants," the spontaneous revolutions
of tendrils and of the stems of climbing plants were

investigated, and the causes producing them reduced to simple laws.

" Insectivorous Plants " was a typical piece of Darwinian work. " In 1860," he says, " I was idling and resting near Hartfield, where two species of *Drosera* abound, and I noticed that numerous insects had been entrapped by the leaves. I carried home some plants, and on giving them some insects, saw the movements of the tentacles, and this made me think it possible that the insects were caught for some special purpose. Fortunately, a crucial test occurred to me, that of placing a large number of leaves in various nitrogenous and non-nitrogenous fluids of equal density ; and as soon as I found that the former alone excited energetic movements, it was obvious that here was a fine new field for investigation."

These researches showed that the plants secreted a digestive fluid like that of animals, and that insects were actually used as food.

The " Power of Movement in Plants," a tough piece of work, was published in 1880. " In accordance with the principles of Evolution it was impossible to account for climbing plants having been developed in so many widely different groups, unless all kinds of plants possess some slight power of movement of an analogous kind."

We come now to the last of his books, a singularly interesting and most characteristic piece of work. On the 1st of November 1837—*i.e.*, about a year after his return from the voyage of the *Beagle*—Darwin read a paper before the Geological Society

on the "Formation of Mould," in which he called
attention to the characters of vegetable mould or
earth; to its homogeneous nature, whatever was the
character of the subsoil; and to the uniform fineness
of its particles.   He gave the history of some fields,
some of which a few years before had been covered
with lime, and others with burnt marl and cinders,
and showed that on cutting into the soil, the cinders
or lime were found, in a fairly uniform layer, some
inches below the turf—as though, to quote the
farmer's opinion, the fragments had worked themselves down.

The explanation he offered, originally suggested
by his uncle, Mr. Josiah Wedgwood, was that earthworms, by bringing earth to the surface in their
castings, must undermine any objects lying on the
surface, and thus cause an apparent sinking.
According to his habit, Darwin continued steadily
to accumulate observations, to devise experiments,
and to collect information from all possible sources.
Finally, in 1881, or forty-four years after his first
paper on the subject was published, his last book—
the "Formation of Vegetable Mould through the
Action of Worms"—was published.   He describes
himself as getting almost foolishly interested in it;
and when the book was published, it was received
with what he describes as "almost laughable
enthusiasm," 8000 copies being sold in about three
months.

Earthworms can live anywhere in a layer of earth,
even if thin, provided it retains moisture; dry air
is fatal to them, but they can live submerged in

water for nearly four months. They live chiefly in superficial mould, from four to ten inches in thickness, but may burrow into the subsoil to a much greater depth. The burrows are effected partly by pushing away the earth on all sides, and partly by swallowing it. In cold weather they may burrow to a depth of from three to eight feet. The burrows do not branch, and are lined by a layer of earth voided by the worm; they end in slightly enlarged chambers in which the worm can turn round.

Earthworms are nocturnal, remaining in the burrows all day and coming out to feed at night. They usually keep their tails in the burrows while feeding, but may leave them entirely, and can crawl backwards or forwards. They are very sensitive to vibrations in the earth, a fact which can be shown by placing a pot of earth containing them on a piano. They do not, however, seem to hear, but can distinguish light from darkness.

Their food consists of leaves or fresh raw meat, especially fat.

The worms drag the leaves into their burrows, using their lip to lay hold of them, and showing judgment as to which end to draw them in by. They swallow enormous quantities of earth, which they void in spiral heaps, forming worm castings. By means of the gizzard the earth is ground and mixed with vegetable matter, and the castings are hence of a black colour. They are in this way always bringing earth from below and depositing it on the surface.

Darwin showed that a layer of coal cinders, spread

over a field, had sunk an inch in about five years, or
one-fifth of an inch per year.   In the case of a stony
field with very scanty vegetation, and covered
thickly with flints, some half the size of a child's
head—all these had disappeared in thirty years.

The weight of the castings, collected and dried,
which were ejected on a square yard of ground on
Leith Hill Common during a year, amounted to over
7 lbs., or 16 tons per acre.

At Stonehenge some of the outer ring of stones
have fallen, and are now partially buried in the soil,
owing to the action of worms.   The specific gravity
of objects does not affect their rate of sinking.   In
fact, "worms have played a more important part in
the history of the world than most persons would at
first suppose."

Hensen placed two worms in a vessel eighteen
inches in diameter, which was filled with sand on
which fallen leaves were strewed : these were soon
dragged into burrows to a depth of three inches.
After about six weeks an almost uniform layer of
sand, nearly half an inch thick, was converted into
mould, from having passed through the alimentary
canals of these two worms.

It is not difficult to account for the success which
this work achieved.   "In the eyes of most men the
earthworm is a mere blind, dumb, senseless, and
unpleasantly slimy annelid.   Mr. Darwin undertakes
to rehabilitate his character, and the earthworm
steps forth at once as an intelligent and beneficent
personage, a worker of vast geological changes, a
planer down of mountain-sides, a friend of man."

DARWIN'S METHOD OF WORK.

On this subject his own estimate and account are most helpful. His method was eminently *inductive*, consisting in the accumulation of large numbers of facts, which he compelled to yield up their secrets. He had the greatest distrust of deductive reasoning, and objected to his grandfather Erasmus Darwin's "Zoonomia," because of the undue proportion of speculation to the facts given. He observes: "From my early youth I have had the strongest desire to understand or explain whatever I observed; that is, to group all facts under some general laws. These causes combined have given me the patience to reflect or ponder for any number of years over any unexplained problem." Professor Huxley notes how "that imperious necessity of seeking for causes, which Nature had laid upon him, impelled, indeed compelled him, to inquire the how and the why of the facts, and their bearing on his general views."

His patience was extraordinary, and the dogged insistence with which he stuck to a subject and followed a clue wherever it led, was wonderful. He spent eleven years over his investigations on orchids; sixteen years over those on the insectivorous plants; the "Origin of Species" was the result of twenty-one years' work; and "Earthworms" that of forty-four years of observations and experiments. He had an unusual power of noticing things which easily escape attention, and of observing them carefully.

" I had," he says, " during many years followed a
golden rule, namely, that whenever a published fact,
a new observation or thought came across me, which
was opposed to my general results, to make a
memorandum of it without fail and at once ; for I
had found by experience that such facts and thoughts
were far more apt to escape from the memory than
favourable ones. Owing to this habit, very few
objections were raised against my views ('Origin of
Species') which I had not at least noticed and
attempted to answer."

" I gained much by my delay in publishing, from
about 1839, when the theory was clearly conceived,
to 1859 ; and I lost nothing by it."

PERSONAL CHARACTERISTICS.

Darwin was about six feet high, of active habit,
but with no natural grace or neatness of movement.
He was awkward with his hands, and unable to
draw at all well. He had a full beard and grey
eyes, overhung by extraordinarily prominent and
bushy eyebrows. In manner, he was bright, ani-
mated, and cheerful ; a delightfully considerate host ;
a man of natural and never-failing courtesy—leading
him to reply at length to letters from anybody, and
sometimes of a most foolish kind. He was fond of
animals, and had a great power of stealing the
affections of other people's pets.

His private life was burdened by almost constant
illness, which rendered him incapable of work for
weeks, or even months at a time ; and in later years

these long periods of suffering often made work of any kind an impossibility. "For nearly forty years he never knew one day of the health of ordinary men, and his life was one long struggle against the weariness and strain of sickness."

Under such conditions absolute regularity of routine was essential, and the day's work was carefully planned out. At his best he had three periods of work : from 8 to 9.30 ; from 10.30 to 12.15 ; and from 4.30 to 6. Each period being under two hours' duration.

Darwin was a man greatly loved and respected by all who knew him. There was a peculiar charm about his manner, a constant deference to others, and a faculty of seeing the best side of everything and everybody.

The striking characteristic of his manner of work was his respect for time. His natural tendency was to use simple methods and few instruments ; little odds and ends were saved for the chance of their proving useful. One quality of mind, which seemed to be of special and extreme advantage in leading him to make discoveries, was the power of never letting exceptions pass unnoticed. He enjoyed experimenting much more than work which only entailed reasoning.

For books he had no respect, regarding them merely as tools to be worked with, and he did not hesitate to cut a heavy book in half, to make it more convenient to hold. He marked the passages bearing on his work, and made an index at the end of the volume. Like many eminent people, he experienced the greatest difficulty in writing intelligible English, and took much pains to accomplish this. "There seems," he says, "to be a sort of fatality in my mind leading me to put at first my statements or propositions in a wrong or awkward form." His tone of writing was courteous and conciliatory, and he deliberately avoided controversy.

The closing scene in Darwin's life was in the early months of the year 1882, when his health underwent a change for the worse, and on the 19th of April he died. On the 24th he was buried in Westminster Abbey, in accordance with the general feeling that such a man should not go to the grave without public recognition of the greatness of his work.

And of that work, how shall we estimate its value? To form any notion, however inadequate, we must try to realise the world into which he was born; to picture to ourselves what naturalists up to his time were doing, and what were their aims, ambitions and methods.

From the time of Linnæus the majority of naturalists were devoted to classifying, naming, and labelling animals, and then leaving them. Others went further and studied more deeply. Increase of knowledge led to constantly increasing specialisation and division of labour; each worker coming to look on his own department as more or less isolated and independent. There was no bond of union between these men, no system, and no true basis of classification.

What Darwin did was to put the backbone into

the whole structure; to knit together knowledge from all sources; to point out clearly what was the real nature of these mysterious affinities between the animals now living; to render possible the conception of Natural History as one coherent whole; to show that the real bond was one of blood-relationship, and that the differences between fossil and living animals, which so impressed Cuvier, were not difficulties in the way of such blood-relationship, but necessary consequences of it; to show men that there was no need for them to invoke mysterious agencies to effect they hardly knew what; to show them that all they had to do was to look about, to follow Lyell's method, and see what was now happening around them, in order to get the clue to the past.

Not merely has he changed the whole aspect of biological science, giving it new aims and new methods; but the influence of his work has spread far beyond its original limits. Principles and laws, first established by him for biology, are now recognised as applying to all departments of science, indeed to all departments of knowledge; and it is to him that the phrase the "Unity of History" owes its real significance.

And if we are struck with the importance and grandeur of the results obtained, so are we equally impressed with the simplicity of the means by which they were achieved. The lesson to be derived from Darwin's life and work cannot be better expressed than as the *cumulative importance of infinitely little things.*

Such was the man whom we revere and marvel at, for the greatness of his services to mankind and his contributions to human knowledge ; and love for the truthfulness, the patient endurance in suffering, and the gentle courtesy of his life.

INDEX

Printed by BALLANTYNE, HANSON & CO.
London and Edinburgh.

www.ingramcontent.com/pod-product-compliance
Lightning Source LLC
Chambersburg PA
CBHW021522210326
41599CB00012B/1351